Studies in Systems, Decision and Control

Volume 625

The series "Studies in Systems, Decision and Control" (SSDC) covers both new developments and advances, as well as the state of the art, in the various areas of broadly perceived systems, decision making and control–quickly, up to date and with a high quality. The intent is to cover the theory, applications, and perspectives on the state of the art and future developments relevant to systems, decision making, control, complex processes and related areas, as embedded in the fields of engineering, computer science, physics, economics, social and life sciences, as well as the paradigms and methodologies behind them. The series contains monographs, textbooks, lecture notes and edited volumes in systems, decision making and control spanning the areas of Cyber-Physical Systems, Autonomous Systems, Sensor Networks, Control Systems, Energy Systems, Automotive Systems, Biological Systems, Vehicular Networking and Connected Vehicles, Aerospace Systems, Automation, Manufacturing, Smart Grids, Nonlinear Systems, Power Systems, Robotics, Social Systems, Economic Systems and other. Of particular value to both the contributors and the readership are the short publication timeframe and the world-wide distribution and exposure which enable both a wide and rapid dissemination of research output.

Indexed by SCOPUS, DBLP, WTI Frankfurt eG, zbMATH, SCImago.

All books published in the series are submitted for consideration in Web of Science.

Maria Isabel Aldinhas Ferreira

Transitions of the Immutable: Reflections on Public Education in the 21st Century

Springer

Maria Isabel Aldinhas Ferreira
Language, Mind and Cognition Group,
Faculdade de Letras
Centre of Philosophy, Institute for Systems
and Robotics—ISR/IST
University of Lisbon
Lisbon, Portugal

ISSN 2198-4182 ISSN 2198-4190 (electronic)
Studies in Systems, Decision and Control
ISBN 978-3-032-02808-2 ISBN 978-3-032-02809-9 (eBook)
https://doi.org/10.1007/978-3-032-02809-9

This Springer imprint is published by the registered company Springer Nature Switzerland AG
The registered company address is: Gewerbestrasse 11, 6330 Cham, Switzerland

To my Students

Pedagogy is relational.
Both teachers and learners are transformed
through the pedagogical encounter
And from this encounter a Better World
should emerge.

Preface

Universal Education is essential to human flourishing, personal fulfillment and an important contribution to individual and collective happiness. It is per se a Human Right, as stated in article 26 of the 1948 Universal Declaration of Human Rights [1], but it is also the pathway towards individual and collective freedom and development. Universal education is the basis of an equitable society where, independently of origin, social provenance, age or physical condition no one is set apart, where everyone has the chance to find and develop their own talents, whatever the stage of life they are at, defining or redefining their own track to self-realization and happiness, while contributing this way to a common more humanly rich social edifice.

The Public School system is the cornerstone of the process of humanization that education should constitute. It is the key that opens the door to self-fulfillment, guaranteeing equity and inclusiveness, fostering peace and harmony in a hopefully cherished healthy planet. As [2: 4, I] points out:

> From early childhood education and care (ECEC) to tertiary and adult education, education and training are the foundation for successful participation in society and one of the most powerful tools for building more inclusive societies.

In the midst of a technological revolution that has been transforming the modes of production, the distribution flows and services, restructuring economies and determining new lifestyles and the very way of being in the world of individuals and societies, schooling emerges as a fundamental time and the privileged setting to develop the essential skills, required to face the challenges of a world object of essential transformations, while acquiring the ethical awareness necessary to handle intelligent tools in a responsible and beneficial way.

In a world undergoing relevant demographic, social, geopolitical and technological changes and on the verge of a civilizational mutation, these brief reflections address what is here called the inherently immutable character of the School as an institution, i.e., its resistance towards change, even when apparently following and cherishing innovation. Casting a critical look on three important aspects of western public schooling's present reality, this text aims to identify some of the pathways

School has to follow to respond to the challenges of a changing world while keeping what should be its main goal: a happy human flourishing.

In the Global Pact to Education, Pope Francis highlights [3]:

> Today, we are called to have the necessary *parrhesía* to leave behind superficial approaches to education and the many shortcuts associated with utility, (standardized) test results, functionality and bureaucracy, which confuse education with instruction and end up atomizing our cultures. Instead, we should aim to impart an integral, participatory and polyhedral culture. [...] The value of our educational practices will be measured not simply by the results of standardized tests, but by the ability to affect the heart of society and to help give birth to a new culture. A different world is possible and we are called to learn how to build it. This will involve every aspect of our humanity, both as individuals and in our communities.

Lisbon, Portugal — Maria Isabel Aldinhas Ferreira

References

1. Universal Declaration of Human Rights: https://www.un.org/en/about-us/universal-declaration-of-human-rights
2. Communication from the commission to the European parliament, the council, the European economic and social committee and the committee of the regions. Action plan on integration and inclusion 2021–2027. https://eur-lex.europa.eu/legal-content/EN/TXT/?uri=CELEX%3A52020DC0758
3. *Message of His Holiness Pope Francis on the occasion of the meeting organised by the congregation for catholic education:* Global compact on education. Together to look beyond. https://www.vatican.va/content/francesco/en/messages/pont-messages/2020/documents/papa-francesco_20201015_videomessaggio-global-compact.html

Acknowledgements

The author wants to address a word of appreciation and gratefulness to Prof Isabel Braga,—Escola D. Antonio da Costa, Almada, Portugal—the most creative and inspired librarian I have ever known, for many hours of hard work and dedication that allowed the realization of many multicultural and educational activities. Many thanks also to D. Senhorinha and D. Deolinda for their assistance during the events that took place in the school's library and to D. Ana Rosa, Sr. Pedro and D. Carla for their technical assistance. To Prof. Helena Pinto I thank the support in the IDILICA project and in the activities aiming at the integration of immigrant students.

To my colleagues Isabel Braga, Helena Pinto, Silvia Macedo, Ana Sofia Fernandes António Marques, Ana Pereira, Francisco Destapado, Celia Vilar, Joana Rosário, Graça Sintra and so many others I am grateful for sharing a vision of the role of school in a world undergoing deep demographic, economic, social, cultural and technological transformations.

To Thomas Ditzinger I thank for his kindness and for supporting the initial idea from its very start and to all the Springer team for their solicitude and extreme professionalism throughout the making of this book.

Contents

Chapter 1
Introduction

I would like to imagine the school in a few years as a space for building individuals, where it would be possible not only to learn theory but also to grow socially, communicatively, affectively, and in character. A space where we would not learn the answers, but rather how to get to them. A space where we could learn with pleasure (English translation from the Portuguese original).
João Roque (student) in newspaper Sol, 12th December 2024

"Transitions of the Immutable: Reflections on Public Education in the 21st Century" is a set of brief reflections—Working Paper #1 (On Learning: The Importance of Respect, Kindness, and Well-being), Working Paper #2 (Immigration, Inclusion, and Linguistic and Cultural Diversity), Working Paper #3 (Technological Revolution, Pedagogical Innovation, and Ethics)—on the constancy of the School throughout times, i.e., its apparent rigidity, the conservatism of its processes as an institution, and the challenges the School has been facing for several decades in a context of increasing globalization, marked by a dynamic of profound demographic, economic, social, and ideological changes; rapid technological development with high disruptive potential, amidst a climate crisis threatening existence on the planet, and on the brink of a new civilizational paradigm [1].

On the need for an educational response to these challenges, [2: 1] states:

> Our world is at a unique juncture in history, characterized by increasingly uncertain and complex trajectories shifting at an unprecedented speed. These sociological, ecological and technological trends are changing education systems, which need to adapt.

This reflection thus assumes an essentially prospective character by aiming to contribute to the necessary educational adaptation to the new demands of this complex evolving context, questioning the paradoxical apparent conflict between the role of the School as an instrument of training/shaping individuals, determined by the demands of a given socio-economic system and in the capitalist context- the market needs, and what should be its true desideratum while responsible for the integral development, personal fulfillment, and well-being of the citizens who attend it from an early age.

M. I. A. Ferreira, *Transitions of the Immutable: Reflections on Public Education in the 21st Century*, Studies in Systems, Decision and Control 625,
https://doi.org/10.1007/978-3-032-02809-9_1

On the role of the school in developing different types of skills, the WEF report (2018:4), [3] dedicated exclusively to learning refers:

> Education should equip students with the skills they need to lead healthy, productive, meaningful lives.

We cannot help but notice, when reading this formulation, the absence of what seems to us should be the main desideratum of the School—to contribute to the construction of Happy Lives. Let's recall the words supposedly attributed to John Lennon [4] on this topic: "When I was 5 years old my mother always told me that happiness was key to life. When I went to school they asked me what I wanted to be when I grew up. I wrote down 'happy'. They told me I didn't understand the assignment, and I told them they didn't understand life."

On this fundamental mission of the School, Noddings [5] emphasizes: "When parents are asked what they want for their children, they usually answer that they want their children to be happy. Why, then, is happiness rarely mentioned as an aim of education?".

The answer to this question is easy to find when we identify the main purposes that seem to guide, for example, the educational systems of the so-called developed countries. Regarding these objectives and purposes, Bell [6: 31] states:

> Education is always implicitly or explicitly a political issue (....). What is taught, what is not taught, how students are taught and how educational institutions are organized are fundamentally political questions. Education cannot be disconnected from wider views about the society in which it is located.

Spring (2011) [7], cited by Bell [6], identifies what he designates as the dimensions of the three main educational purposes in developed Western societies—the political, social, and economic. According to Spring, the political purpose of education is to enable young people to become full-fledged citizens, active participants in a liberal democratic system. The social aspect of education aims to reduce inequalities, favoring conditions of equity that guarantee the development of all citizens. The economic purpose aims at the growth and maintenance of the workforce necessary for a given socio-economic model that currently transcends mere national or regional boundaries and increasingly assumes an international character. Capital/the markets require quality (trained) and sufficiently large workforce for the functioning of its structures. It is, therefore, up to the educational systems, to the School, to ensure the response to these needs.

If the production model sustaining a society is the dynamo that determines its social structuring and defines its cultural and political substance, it also determines the conception and design of the educational model to be followed in that society. This fact results, whether one likes to admit it or not, in a utilitarian view of the School's role that in capitalist societies prioritizes market needs and is particularly reductionist concerning the conditions for the talents' development, the realization and happiness of particular individuals.

Although it is not possible to dissociate the School from the economic, social, and cultural setting it emerges from and from the demands required by the maintenance and/or evolution of that setting, its true vocation should always be to promote the integral development of the human beings who attend it. By corresponding to their talents and interests fostering the flourishing of their multiple capacities, favoring the development of cognitive, emotional, social…competences that will allow them to find or create the conditions in the community to be integrated, grow as citizens, fulfill themselves as persons and be happy. As the Convention on the Rights of the Child (UN, 1989) [8] highlights:

> the education of the child shall be directed to the development of the child's personality, talents and mental and physical abilities to their fullest potential.

It is up to governance to ensure the right to this development to all citizens through adequate Public School policies, that grant the conditions of equity, personal flourishing that guarantee a more prosperous and developed society and a fairer and more balanced social context. Reimagining Future Together [2: 11] refers to the need to reflect on the limits of the present policies on education and the urge to construct a new social contract capable of answering the challenges of today's world:

> Constructing a new social contract means exploring how established ways of thinking about education, knowledge and learning inhibit us from opening new paths and moving towards the futures we desire. Merely expanding the current educational development model is not a viable route forward. Our difficulties are not only the result of limited resources and means. Our challenges also stem from why and how we educate and the ways we organize learning.

The question of whether there is a true correspondence between the duration of compulsory schooling and the actual development and consolidation of the competences necessary for personal development is fundamental. This issue is particularly pressing in the educational contexts of developed countries where long-term school attendance is compulsory. In the case of Portugal this attendance is, since 2009 [9], until the 12th grade or until the age limit of 18 years old.

But as the WEF [3: 5] states:

> Schooling without learning is a terrible waste of precious resources and of human potential.

And it is here that the Public School, which should be the primary promoter of individual and collective development and freedom by equipping all individuals who attend it with the necessary skills and competences to thrive and succeed, often fails its mission. Following the same rigid standardized learning route designed for an abstract generality and ignoring individual characteristics simply serves to burn time, foster students' frustration and promote vice and violence. As Azevedo [10] (p. 145) highlights in this regard:

> Declaring the obligation to attend school until the age of 18 is equivalent to establishing a political commitment [...] around the equity and quality of the education that the school provides to each specific citizen [....] having as a reference the person who lives in each student. Otherwise, this declaration becomes mere state violence over citizens.

Compulsory schooling must, therefore, be a productive time for growth, for finding out the joy that results from the magic of learning, and also a time for discovering oneself, one's talents, and one's possible place in the world. It should never be a painful physical and mental drag through a rigid sequence of predefined contents, some of them unquestioned and unchallenged for decades, for which, so often, neither usefulness nor interest can be foreseen, distributed over precious years of life, thus ingloriously wasted.

The School cannot continue to be a source of frustration for many, of permanent confrontation with the apparent inability to meet standards, metrics that have managed to resist all movements of pedagogical innovation and which inflexibly persist to this day and are defined abstractly for a hollow generality.

On this topic, a student interviewed by the Portuguese newspaper Sol, on the 12th December 2024, referred:

> When I get up every day, I go to school to learn. The problem is that, often, the school becomes a space centered on another word: assessment, such as tests or exams. Whether because they consider this method effective in promoting learning, or because they do not have, or do not want to have, another way to assess students' abilities. I find it regrettable that evaluation is prioritized over learning.

The School cannot continue to be, for some, little more than serving a sentence of about 12 years of prison, as Memories from the Battlefield [11] very well illustrated. In this documentary, we can see how it is incredibly possible for a child to enter the 5th grade at the age of 10 and exit school eight inglorious years later, without having even completed the 6th grade (2nd cycle). Having uselessly consumed all that time of their lives, not obtaining any type of training that will be useful or illuminates the possible path ahead.

Therefore, it is essential that the learning process exhibits a plasticity that allows its adequacy to individual subjects and not just target abstract masses. It is also essential that it takes place in an environment conducive to holistic individual development, grounded in deep ethical and humanistic values where respect for the student as a person and as a citizen, concern for their physical and mental well-being, and kindness are fundamental factors.

It is also essential that the formative role of the School is not restricted to the needs of the productive fabric, of a "market", at a given moment in the historical narrative, nor determined by the interests of any type of corporation but allows each individual to discover what truly interests them, what they would like to dedicate themselves to. It is, therefore, unacceptable, for example, the frequent insistence, through various campaigns, on choosing paths that privilege STEM relegating to a subsidiary place, to a niche, the so-called Humanities, namely Classical Languages, History, Philosophy as well as the Arts in general. It is also unacceptable to artificially create numerus clausus, that dramatically limit access to many higher education courses, thus preventing thousands of students from fulfilling what is felt as a vocation, what is seen as a dream. The result of this incorrect perspective on the role of the School creates undesirable imbalances, both at the individual and collective level, which are already quite visible in the glaring shortage of professionals in fundamental areas. Finally, it is also unacceptable the political/economic emphasis on developing in students a spirit of "entrepreneurship" that stifles the rich and varied creative dimensions of human ingenuity, focusing exclusively on the objective of "making money".

In the present reflection, an attempt was made to account for the immutability of the School as an institution over decades, its iron resistance to change, despite all the movements of pedagogical and technological innovation, which it has gradually been incorporating over time, for brief moments, but which it has simultaneously, progressively been erasing, discarding, until almost returning to the primordial initial model generated by the industrial revolution. As Azevedo (ibid: 222) states:

> Most of these innovations are tolerated, accepted, and even developed, like the fresh water of legitimation, which is slowly absorbed and dissolved by the corrosive liquid of ambiguity until the day it ends up being devoured as nutritious food so that this great unstable and ambiguous equilibrium remains active and current....

And why is it so? Because all fundamental institutions, and the School is one of them, are necessarily conservative. This conservatism stems from their importance in maintaining a certain socio-economic and cultural model, whose stability and continuity they guarantee. As Daniel Hager [12] stated:

> The public sector tends to put a premium on conservatism. Bureaucratic organization seeks self-perpetuation. Bureaucracies rarely relinquish status through their own volition.

All reforms, even punctual ones, in this complex organism that is the Public School system have individual and collective impacts. In order not to alter what is considered the essential format, only ad hoc measures generally dictated by trends are accepted. And these, like all fashions, are ephemeral and consequently temporary.

In this context of immutability that characterizes the School as an institution, there is a glorious exception. An exception that bursts forth as a truly Explosive, Alive, Liberating, and Regenerative force: the elimination of Fear from the relationship with the institution and the elements of "power" that substantiate it.

Fear, which was unfortunately present in the School in the past, in more conservative school contexts, all over the world, and which prevailed in the Portuguese School until the Revolution of April 25, 1974. This fear was fostered by cynical cold austerity, by the animosity of the institution that thus exercised control and dominance over its young citizens. This fear, experienced by students, of making mistakes, of doing something wrong and being consequently severely punished, this fear that crippled children and young people, that gagged them, that poisoned the learning process by embittering it with anxiety, this fear that destroyed by suffocating the free flourishing of young individuals, this Fear has disappeared, it no longer exists!!!!! And, in spite of all the radicalisms that are emerging today in our societies, this corruptive Fear cannot be allowed to install itself again, because it is destructive, poisoning and shameful. And as this poisoning fear is not allowed to emerge, also all forms of anxiety and mental stress produced within the school's walls should be eliminated, namely those nurtured by the testing based on standardized established metrics of "successful performance". As Reimagining the Future refers (page 54).

> All educational environments should be places of safety, even of refuge, where learners are encouraged to experiment, dare, fail, and create. Pedagogy should stimulate imagination and creative thinking, and promote intellectual freedom, which includes the right to make mistakes and learn from them.

In this context, the Public School has today, despite its essential immutability, a horizon of hope if it knows how to fulfill, in the twenty-first century, its transformative function: **teaching how to learn**, doing so with respect and kindness for the individuals who attend it, regardless of their age, origin, or social status. Fostering intellectual freedom, critical thinking and nurturing the quest driven by the passion for learning. Teaching the learners of the twenty-first century to stand on their own, critically and creatively, learning and evolving in a world where information, interaction, and creation can be now just a click away. This teaching of how to learn will equip everyone with the ability to, with creativity, but also with a critical spirit, seek answers, design strategies, and, throughout life, face the challenges of a changing world, designing life paths that are desired to be happy.

References

1. Ferreira, M.I.A. (2020) *How Smart Is Your City? Technological Innovation, Ethics and Inclusiveness*, Maria Isabel Aldinhas Ferreira (ed), Springer, Cham, 2020. ISBN: 978-3-030-56925-9. https://doi.org/10.1007/978-3-030-56926-6. https://unesdoc.unesco.org/ark:/48223/pf0000379707
2. Reimagining Our Futures Together: A New Social Contract for Education, UNESCO Report 2021. International Commission on the Futures of Education, ISBN 978-92-3-100478-0 https://doi.org/10.54675/ASRB4722. https://unesdoc.unesco.org/ark:/48223/pf0000379707
3. Learning to Realize Education's Promise—World Development Report (2018) https://www.worldbank.org/en/publication/wdr2018
4. John Lennon Quote https://www.goodreads.com/quotes/282517-when-i-was-5-years-old-my-mother-always-told
5. Noddings, N. (2003), (2009) Happiness and Education**.** Cambridge University Press. https://doi.org/10.1017/CBO978051149992
6. Bell, L.A. (2020). Education Policy: Development and Enactment—The Case of Human Capital. In: Fan, G., Popkewitz, T.S. (eds) Handbook of Education Policy Studies. Springer, Singapore. https://doi.org/10.1007/978-981-13-8347-2_2
7. Spring, J. (2011). *American education.* New York: McGraw-Hill.
8. Convention of the Rights of the Child (UN, 1989) https://www.unicef.org/child-rights-convention
9. https://diariodarepublica.pt/dr/detalhe/lei/85-2009-488826
10. Azevedo, J. (2024) Modo de Produção da Exclusão Escolar. Olhar a escola a partir dos excluídos. Fundação Manuel Leão
11. Memórias do Campo de Batalha (2005), A documentary directed by Isabel Ferreira, in which interviews conducted by the teacher, a group of young people from an alternative curriculum class at Escola Básica 2, 3 in Almada, reflected on themselves and their relationship with the school.
12. Daniel Hager Henry Ford's (1927) Reinvention and the Implications for Public Education. December 11, 2002 https://www.mackinac.org/4926

Chapter 2
Working Paper #1 on Learning: The Importance of Respect, Well-Being and Kindness

1 Epistemological Assumptions

1.1 The Child: Person and Citizen

A large part of the legal frameworks currently in force in developed Western societies determine that an individual has the legal status of child until the age of 18:

> [...] a child means every human being below the age of eighteen years unless under the law applicable to the child, majority is attained earlier. [1]

These legal frameworks reflect the consensual notion that human beings only reach their full biological, emotional and social autonomy at a late stage of their development, compared to other species. It is also commonly acknowledged that this development should take place in the protective family environment, capable of providing and summoning the resources necessary to a harmonious physical, emotional and social maturation process from which will hopefully emerge a responsible and socially committed adult citizen. These facts are responsible for the prevalence of a paternalistic attitude towards children by society, in general, and by its main institutions in particular. As Cohen [2] states, the ideology he designates as the "caretaker," already defended by John Locke in his Second Treatise On Government [3], considers that the main role of parents is that of caregivers of their children, considering that they are not yet complete human beings. According to Cohen [ibidem], this perspective, which to some extent has prevailed until today, assumes that it is the adults who know what is best for the child, acting consequently in their best interest. Obscuring the possibility of conflict of interests between the child and the adult, this perspective favors a paternalism that, according to that author, undermines the dignity of the child by ignoring fundamental rights.

This attitude, which in abstract may seem protective and defensive of the child's best interests, is, in fact, responsible for the conception of the child as an entity that, for not having reached a state of maturity and social autonomy—the "age of reason",

M. I. A. Ferreira, *Transitions of the Immutable: Reflections on Public Education in the 21st Century*, Studies in Systems, Decision and Control 625,
https://doi.org/10.1007/978-3-032-02809-9_2

i.e., the time when individuals are capable of weighing the pros and cons of their decision—does not yet possess exactly the same rights as adults, particularly with regard to the respect their person demands in terms of treatment and the respect for their will and opinion.

Nishiyama [4] points out that traditionally, children are considered as mere *future* citizens who are acknowledged only when they behave in accordance with adults' expectations. In this sense, according to this author, children are one of the last frontiers of democratic inclusion, as they have long been seen but not heard.

As Tamar Shapiro states [5: 1]:

> Implicit in our moral and legal practice is the idea that children are persons, but not full ones. [...] which means two things: first, they are not permitted to exercise the full range of rights enjoyed by adults, and second, they do not carry the full burden of adult responsibility. As for rights, although children have an equal right to have their basic interests protected and promoted, they do not have an equal voice in deciding matters that bear on their interests. In particular, children do not have the prerogative to waive their right to be protected and helped by adults; indeed they have to be raised, whether they like it or not. [...] children, on the other hand, can be compelled to go to school, live with their parents, and take nightly baths, purportedly just because they are children, and because these measures are for their own good. Moreover, insofar as children do have some say in how their good is to be promoted, their voices are accorded at most 'consultative,' and not 'authoritative' force. In other words, children's opinions about what ought to be done for them are regarded as merely provisional conclusions, proposals which stand in need of final ratification by a higher, adult authority.

Social maturity and the consequent autonomy that comes from it to make choices, make decisions, and assume responsibility for them, the possibility of full citizenship, results from a maturation process that starts in the protective and potentially educative environment of the family and necessarily continues in the formative environment of the school. But this maturation that the educational institution should promote is not favored, on the contrary, it is eroded [5] by the arbitrary imposition of adult authority and by disciplinary measures when the established rules are not correctly addressed. And in this context, one can ask where in practice is the inalienable right to dignity translated in the respect for the child's person and where is their opinion enshrined?

When contrasting the modus operandi of the school institution, in the past and in the present, an important evolution can be noticed—the abolition of physical punishment in case of rule infringement. A practice that existed for decades as a "disciplinary measure" concomitantly with the exercise of an overwhelming fear-based authority. However, despite this profound evolution and regardless of the established codes of ethics [9] that enshrine the rights and duties of students at school, we find that in daily practice there is still a long way to go in terms of respect, deference, and kindness owed to the child's person.

As Cohen [2] defends, being a person, the child is entitled to have as many rights as adults do, and these rights should be observed in daily practice and interaction with older people. Respected by parents and respected by all those who interact with them in different contexts, particularly those responsible for their formal or informal education, i.e., educational aids, educators, and teachers. The claim that a child, a young person, is not just a "citizen to be" but a person and a full-fledged citizen (it should be noted that the citizen card, issued at birth, not only grants them a legal and juridical identity but also the necessary status to enjoy that citizenship) is a transformative and regenerative assumption with enormous consequences also in the way the school as institution relates to their young learners whose cognitive, social, and emotional development it should favor while respecting their individuality.

As the Sustainable Development Goals, target 4.7, point out, students are global citizens who require the knowledge and competencies to build sustainable futures in an increasingly interdependent world.

It's the school's mission to provide the tools to achieve this.

1.2 Positive Emotions Enhance Successful Performance

The British Council had, years ago, in the brochure publicizing their language courses, an exemplary phrase.

This simple, seemingly trivial phrase highlights the importance of a state of satisfaction, of physical and mental well-being of the student for a successful learning process (Fig. 1).

The present reflection follows Lyubomirsky et al. [6: 803] in the definition of "happy people"—"[…] those who experience a preponderance of positive emotions

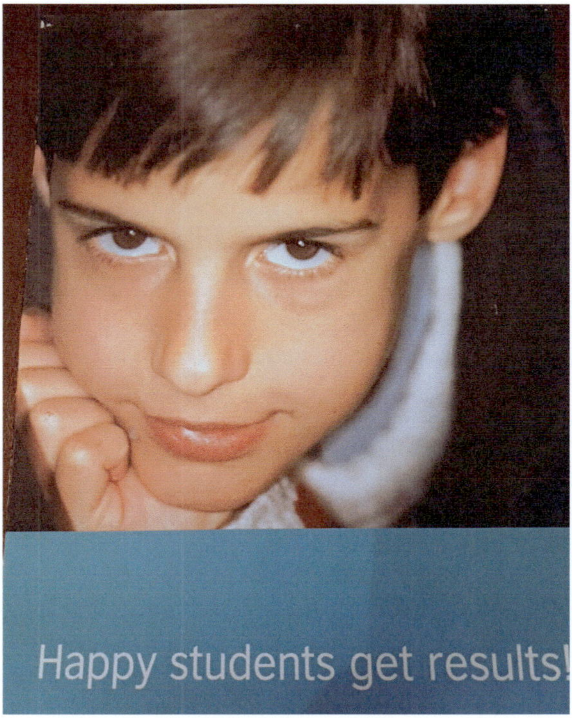

Fig. 1 Happy students get results (credits to the British Council—British Institute)

over time" and in the assumption of the eventual positive effect of this internal state in terms of individual performance [ibid: 804].

> Positive valenced moods and emotions lead people to think, feel, and act in ways that promote both resource building and involvement [...] Positive emotions produce the tendency to approach rather than to avoid and to prepare the individual to seek out and undertake new goals.

In their research, Lyubomirsky et al. [6, 7] demonstrated that in humans "success" is often associated with the prevalence of positive emotions, but that their co-occurrence should not necessarily be interpreted as a cause-effect relationship, like Success-Happiness, but rather as a result: Happiness—Success, as this is favored by the experience of positive emotional states.

We will not refer to "individual success" given that this term, as those authors also recognize, embodies a series of traits that are culturally determined:

"Being successful means accomplishing those things that are valued by one's culture, flourishing in terms of the goals put forth by one's society." [6: 822].

Instead, the expression "successful performance" will be used in this text, as being that which corresponds to achieving an objective. In the school context, this performance may correspond to the quality of the simple execution of a classroom task according to established guidelines and stages, the correct completion of a specific project, or the achievement of a significant milestone in terms of the school path, such as a final exam of a cycle.

2 Creating Favorable Learning Frameworks

2.1 Allardt's Model

In order to identify the main subjective conditions (inherent to the individual) and the set of objective conditions (related to external factors) that can constraint the "preponderance of positive emotions" by the student and influence their learning process, we will use the model proposed by Allardt [8] in his sociological theory for the analysis of the Welfare State, and whose productivity in analyzing school contexts was demonstrated by the works of Konu and Rimplea [9].

Allardt [ibidem] argues that there are material and non-material conditions essential for the existence and development of the human being. He groups them into three major categories:

1. [Having]
2. [Loving]
3. [Being]

According to Allardt, [Having] refers to all the material conditions necessary for subsistence and existence itself. In this category, Allardt includes firstly the health conditions that are determinants for the life and autonomy of the individual and secondly all the conditions responsible for dignified existence: housing, education, etc.

[Loving] points to the fact that human beings are intrinsically social and need to establish relationships and bonds of affection with others to form their own personal and social identity and for their physical and emotional equilibrium.

[Being] refers to the importance of assuming a role in society—a social identity-, by integrating it, and with their action (work) contributing to its development and simultaneously to their own development as a person. The relevance of their action, of their contribution, is reflected in how their work, their intervention is recognized, valued, and each individual is respected as member of that society.

Recognizing the relevance of these categories, Konu and Rimpelä [9] based themselves on Allardt's model to identify the variables that can determine or condition what they designate as the student's well-being in the school context. The model proposed in [9] considers the categories previously defined by Allardt, adding the category Health.

- **Having**: refers to the conditions of the school. These conditions include the external environment in which the school is integrated and also the internal environment. Regarding the internal environment, Konu et al., [ibid.] distinguish (i) structural/logistical issues, such as lighting, ventilation, comfort, etc., (ii) issues related to the learning environment—curriculum, schedules, number of students per class, etc., (iii) issues related to the services provided, namely meals and health care.
- **Loving**: the social relationships established in the school context, namely the importance of the teacher-student relationship, peer relationships, group dynamics. The impact of negative relationships such as destructive effect of bullying by peers or by adults (teachers or others).
- **Being**: the means provided by the school for personal fulfillment. In this context, each student should be considered an active member of the community, whose opinion should always be heard and taken into account in all matters that concern them or affect them in any way. Regarding learning, individualized teaching and positive feedback are considered fundamental factors for development and a positive emotional state.
- **Health**: the physical and mental health of the student at the beginning of their school journey and throughout it (ages 3–18).

The proposal by Konu and Rimpelä [ibid] demonstrates the productivity of the model conceived by Allardt when applied to assess how school's conditions determine the positive mood and constrain the student's performance. Although greater visibility was given to the variable [Health], which in Allardt's model is contained in the category [Having], Konu and Rimpelä's proposal follows the same fundamental idea: the identification of variables responsible for the individual's well-being and positive emotions in different contexts of social interaction, whether they relate to the dynamics of work or school.

In the next section, we will maintain the premises and categories as defined by Allardt [8], also taking into account the results advanced in [9] regarding the application of the model to the school context.

2.2 [Having], [Loving] and [Being] at School

2.2.1 From Objective Circumstances to Subjective Conditions

Each child is a unique human being, bearing an essential individuality that is translated by an unrepeatable physical reality and a singular psychological and emotional nature that results, on the one hand, from innate personality traits and, on the other hand, from how these innate traits have been defined and developed within the family and social environment (extended family, circle of friends/neighbors, and any educational context prior to compulsory schooling) with which this personality has interacted over the years [10, 11].

These subjective circumstances are the [Having] that each child carries within themselves on arriving at school and which the school (educators, teachers) must at least minimally know to better interact and more easily foster the learning process.

In the set of objective circumstances (external to them), we include those generated by the family context whose affective support—love, attention, and care—is crucial for the emotional balance of the child/young person, for their overall health and harmonious development, favoring or conditioning the evolution of the formal learning process.

We must, however, emphasize that the boundary between subjective and objective conditions is very fluid. Indeed, many objective circumstances (external) exert a profound influence on subjective conditions—their psychological and/or emotional state—their happiness—as it is the case of family disruption in the circumstance of divorce, physical or psychological absence of progenitor or progenitors, family health or economic problems. All these external factors have a high impact on the emotional state of the young learners, frequently causing depression or demotivation and eventually leading to disruptive behaviors compromising their performance. The negative impacts of disturbing circumstances on an adult's psychological equilibrium and performance are easily acknowledged and considered by society, triggering remedial strategies—medical help, psychological help..... to alleviate a negative mind set and try to restore equilibrium. However, in the case of young students, the individual's mood and performance is frequently minimized by the institution, as if by the simple fact of being very young, individuals were easily capable of forgetting, overcoming or even ignoring the psychological pain and distress caused by hard life circumstances.

2.2.2 [Having]: The School's Structural/ Logistical Conditions

We will now briefly approach the set of objective conditions provided by the school environment, following closely the distinctions proposed in [9].

Built forms are endowed with an individualizing significance that results from the "function/role" they are supposed to perform and that pre-exists to the object's conception, determining its specific physical configuration and being responsible for its definition and ontological status [12]. The feature "Function/Role" is responsible for the qualitative distinctness that allows distinguishing a [park] from a [stadium] or a [school] from a [church]. Public and domestic spaces have always been defined according to the different actions, to the routines people will perform there and by the way these routines have been aesthetically translated by different architectural tendencies.

Traditionally, the school building is a space configured for the existence (not necessarily coexistence) of two main types of population: the students and the adults: teachers and the staff. In the typically very rigid configuration of this space, taking away the classrooms, where both categories coexist, there are often zones restricted to students (teacher and staff rooms) and zones that are almost exclusively for the use of students and not usually frequented by adults, such as the student lounge, cafeteria, and often the yards and gardens. This sacralization of space according to the "social categories" present and the functionalities planned for the spaces remain unaltered in most schools throughout times and is also a reflection of the institution's rigidity and immutability over decades, despite various implemented pedagogical experiments.

It is estimated that a significant number of Portuguese school buildings were constructed in the decades between 1960 and 1980, when the number of students increased substantially due to demographics and the extension of compulsory schooling. In 2020, about 40% of Portuguese public schools were more than 30 years old, although this number may vary depending on the region and the type of school: basic or secondary. Despite the Portuguese government having launched the School Park Modernization Program for Secondary Education in 2006 [13], as part of an initiative to improve energy efficiency, ensure accessibility, and adapt educational spaces to modern pedagogical needs, most school buildings still do not reflect nowadays essential architectural and structural/functional concepts such as ergonomic adequacy, inclusive accessibility, or comfort.

Among the structural circumstances that can influence positive mental representations and well-being at school, we identify:

1. The design and configuration of spaces and built forms.
    1.1. The suitability of its architecture to the ergonomics and cognitive and recreational needs of the age group attending it
    1.2. The accessibility for all students, regardless of their physical condition
    1.3. Hygiene and sanitary conditions
    1.4. Ergonomics of equipment (interior and exterior furniture) and computers
    1.5. Comfort

1.6. Beauty

The issue of building/user scale is particularly relevant for children up to 10/11 years old. Take, for example, the case of a school built in the suburban area of Lisbon, in the 1960s and primarily intended for K-12 s, Figs. 2 and 3.

At that time of its construction, the school's plan adhered to all the requirements that defined pedagogical modernity. Large spaces, specific rooms for laboratory activities, specific rooms for visual arts and music, a large gym, and vast sports fields for football, basketball, and handball. A large cafeteria and a library. The exterior area around the school, 2000 m^2, ensured a healthy space for play and recreation.

Fig. 2 The built body (front view) in the end of the 60's

Fig. 3 The built body (lateral view) in the end of the 60's

However, when looking at its functioning nowadays one cannot help noticing a certain degree of "aggressiveness" of that space towards its small users and conclude how the issues of scale, navigation in a large space by a 10-year-old child, and accessibility for all were not even considered for most architectural trends at the time. It was simply assumed that all children would get used to navigate the large space easily and that there would be no particular effort involved in going up and down the four storey building several times a day to move to the laboratory class, or to the music class, or to the visual arts class, or to move to and from the physical education class or simply to go to the exterior space and return from it in the short 10-min intervals between classes. The negative impact of this movement of going upstairs and downstairs, moving to specific classrooms, carrying backpacks almost as heavy as themselves, in an effort that becomes painful by the end of the day, was certainly not anticipated, neither was the issue of scale.

In addition to the topics related to the basic architectural options just mentioned it can be observed in practically all schools that comfort was not, is not, and it is still far from being a concern. Being comfortable is part of a degree of humanization that is missing at schools. Perhaps because for the dominant mentality since the industrial revolution, learning must take place in a somewhat austere environment so that there are no sources of distraction, nor comfort that could make its users excessively relaxed and concomitantly less "hardworking". Given that in most public schools there is no air conditioning, the classrooms are cold in winter and hot in summer, with both sensations being more pronounced in interior regions.

At school, walls are traditionally bare and painted white (though they may eventually later be covered with colorful works done by students), the furniture that does not take ergonomics into consideration, does not adjust to size or weight variations requiring a rigid posture that often becomes painful after a few hours.

In addition to these aspects, there are two spaces in the school that particularly reflect the attention given (or the lack thereof) to the well-being and dignity of young users. These spaces are the canteen and the restrooms/bathrooms.

For a child who, due to social formatting and work practices imposed by the economic framework, regrettably spends practically the whole day at an institution, since infancy up to the end of adolescence, the quality of the food provided and the harmony of the space where the meal takes place are fundamental. The school must also be the site where basic social interaction rules are learned when they have not yet been incorporated in the family context. Sitting properly at the table following the culturally accepted social conventions (such as how to eat using adequately the utensils provided) is as essential as any other know—how that can be learned at school. But for everything to go smoothly, the meal served must meet quality criteria and have a good presentation. We can hardly request correct behavior in terms of posture if what is served is perceived, by taste and/or visual appearance, as not so good.

A critical issue in the students life at school is the status of bathrooms. Their look, status and maintenance reflect the degree of civility of those who frequent them and the care that those responsible for managing these spaces have in preserving an area that is fundamental from a health and hygiene point of view. It is a strong indication that something is very wrong when potential users avoid using them as much as possible. Contrary to what usually happens, bathrooms should (i) have sub-zones perfectly separated to ensure the necessary individual privacy, since childhood (ii) be equipped with all the stuff and equipment necessary for good hygiene and personal care—toilet paper, water, sanitary pads, hand gel, and hand dryer, wall mirrors.

To end this section, we would like to highlight the importance of the presence of beauty—either natural or human made-in an educational environment.

On this topic, we cannot fail to mention, the design of the Munkegaard School (1957) by Arne Jacobsen addressing the relationship between aesthetics, learning and physical well-being. In the design of the Munkegaard School functionality, beauty, and comfort were intentionally favored by the choices made regarding the building and all the interior equipment [14] (Figs. 4 and 5).

An essential humanizing factor, beauty is a very relevant inspiring and harmony-promoting element. It can be present in various forms, from the works created by students in the arts classes, to works by artists exhibited at the school, to the beauty of the surrounding natural forms with which the built form interacts—the beautiful leafy trees of the city, the simplicity of flowers blooming in the middle of nowhere,

Fig. 4 The Munkegaard School Jacobsen's simultaneous focus on aesthetics, learning and physical well-being. https://arnejacobsen.com/works/munkegaard-school/

Fig. 5 The Munkegaard School: classroom. https://arnejacobsen.com/works/munkegaard-school/

the seagulls flying on winter mornings, or the blackbirds hopping over the grass, beauty is there in the small things, children that learn to spot natural in small things and enjoy Nature in all its manifestations will be adults that respect and preserve life in the planet.

2.2.3 [Having]: The School's Curriculum

The curriculum composition and its organization according to a timescale that is supposed to carry students to an expected level of proficiency, respecting pre-determined milestones, defines the rigid backbone of the typical present educational systems. This backbone is a solid tree-like organization that students escalate from a basis up to the top, by accumulating fixed knowledge contents and by complying with standardized performance metrics. Adjusting this backbone or the particular syllabuses that compose it to the interests or specific talents of the child is something, generally, out of the question. In Portugal from the 1st cycle (4 years) to the end of the 2nd and 3rd cycles (5 years), students will follow the same standardized plan in a massified and massifying education system. Regarding this curricular rigidity, Azevedo [15: 178] states:

> [....] an organization based, among other aspects, on teaching by rigidly defined subjects with their hierarchy and incorporating a standardized way of defining and applying national programs, measuring and valuing a very limited set of human knowledge and skills, in a rigid and uniform curriculum management model, class and schedule organization, in a fixed and closed scheme of organizing the teaching/learning space, in a mode of organizing students

> in fixed classes and the cult of hierarchical and often authoritarian relationships between teachers and students[....][1]

The curricular grid in the 2nd cycle (5th and 6th grades) and the 3rd cycle (7th, 8th, and 9th grades) is perceived as being essentially unchanged for decades.

On this topic, in an interview to the newspaper Sol, a student referred [16]:

> Although there has been some evolution, the way teaching and assessment are carried out remains unchanged. Much of what we learn is not essential for our future lives, and many truly useful pieces of knowledge are not taught in school [...] Of course, it is important that any individual has general knowledge of various domains. However, school programs, as they are structured, prioritize very specific and theoretical knowledge over something more practical and useful.[2]

The number of subjects per year is excessive, with an average of 8 subjects annually, concurrently, between ages 10–15. This number decreases when opting for a specific learning area in the 10th grade.

Progress means having successful performance in at least 6 subjects out of the eight-block, and "non-successful performance" cannot occur simultaneously in the native language Portuguese and mathematics. The fact that the different syllabuses, despite of being individually assessed, are in terms of transition considered as a unit, often results in repeating the same disciplinary block for years, frustratingly repeating the same content ad nauseam, even in subjects where satisfactory performance was once already achieved, and learning was successful. This fact is responsible for saturation, disinterest, generalized failure, and even rejection of many students toward school, where they must remain, whether they have successful performances or not, until the age of 18.

With a heavy curricular grid, the schedules are also heavy, with students spending practically the entire day at school. This, however, is something most parents end up appreciating, with school also becoming, due to the demands of a work organization that does not consider the best interest of the child and sidelines or ignores their rights, a sort of "warehouse" where children are left in the early morning (some at 7:45 am) and are picked up in the late afternoon (6:00 pm or even later).

While the structural and functional context provided by the school in the category [Having] always influences, in some way, the student's positive perception of the school environment and their satisfaction in attending it, it is the category [Loving] that exerts a determining influence on the learning process.

2.2.4 [Loving]—The Value of Kindness

"Knowing is intimately connected with feeling. Human intelligence is directly connected with consciousness and affect", as Reimagining the Future (page: 54) highlights.

---

[1] English translation from the Portuguese original.

[2] English translation from the Portuguese original.

The recollection of experiences lived in childhood and youth, in the time and spaces dedicated to learning, persists clearly in the memory of all adults. However, this memory is not often felt as particularly pleasant or gratifying. According to a very small poll informally taken during a gathering with friends (N = 10) (Male = 6, Female = 4) (Ages: 30–65), the answer to the question:

(a) Was your school experience as a learner enjoyable?

was valued with two stars (out of a max of five) by 8 respondents, one respondent assigned it 5 stars, and another one 4 stars.

In the open question "Reasons for disliking the experience", were referred the following causes:

- bullying from peers—3
- animosity from peers—1
- teacher's bullying—2
- teachers' aggressiveness—1
- school's lousy facilities—1

This very small sample shows us that the strongest memories were those related to socio-affective relationships that occurred within the school context involving either adults or peers.

A school can be structurally poor or rudimentary and yet have a warm social environment where students feel welcomed, recognized in their individuality, and appreciated and where, despite the scarcity of resources, they enjoy learning. In fact, the essence of the school, its dynamo, is not the building nor the equipment, although both are very important, but the human warmth that radiates from it, its capacity to welcome, integrate, and love, in a true educational context. In this sense, it is particularly relevant how adults (teachers, staff, and other responsible adults) position themselves towards the person of the Other—the Student—in every way equal to the adult in front of them, except in age, and therefore deserving the same respect, attention and kindness. This respect for their person does not allow for less polite speeches, aggressive words, or any form of roughness. The relationships between all elements of the educational community should be guided by this respect, which is expressed in kindness in speech and action, and by transparency of procedures and attitudes.

Kindness is an essential emotional and social attitude that should be fostered (if not already in the family circle) at school and also developed within it. Conducting a quick search using Google Chrome while writing this text, with the search prompt "the role of kindness at school," we recorded about 295 entries directly addressing the topic [17]

Choosing one of these entries at random: "16 reasons why kindness is important at school" [18]

we can read:

> Teaching kindness is vital to developing children's social and emotional faculties.

Followed by a list of the 16 reasons to do so and some activity suggestions on how to develop this kind of attitude in students.

What stands out from reading the articles associated with those two hundred entries is the emphasis on the need to develop a learning process that leads to the incorporation of the concept to be actualized in specific social behaviors. According to the strategies suggested, this can be achieved by reading short stories or visioning of films, through practical arts work but **nothing** is referred on how this same kindness should be manifested, in the first instance and consistently, in the behavior of the adults (school staff responsible for various sectors and teachers). And it is precisely these adults who are role models that provide the behavioral patterns that will be followed. If a teacher or other staff member addresses students in a dry or harsh manner, they cannot expect the students to respond with kindness. In fact, although the concept of [being kind] does not constitute a symmetrical relationship of the type.

(1) X is married to Y
(2) Y is married to X

that is to say, the fact that

(3) X is kind to Y

does not necessarily imply that

(4) Y is kind to X

it certainly encourages reciprocity because, those who are treated with kindness generally respond in a similar tone or try to erase or diminish the traces of aggressiveness in their response to the Other.

It is interesting to note that the concept [amável], the Portuguese equivalent to [kind] incorporates a semantic duality by accounting simultaneously for the particular favorable attitude of the speaker towards their recipient and, on the other hand, reflecting how this initial positive attitude will make the speaker more pleasant/likable in the eyes of the recipient, as demonstrated by the definition of [amável] provided online by the Oxford Dictionaries[3]:

amável

1. showing delicacy, courtesy; affable, flattering, kind.
2. deserving affection, love; worthy of being loved.

We could represent this syntactic and semantic relationship in the following argument structures:

1. X is kind to Y = X é amável para Y
2. X is worthy being loved, being treated with kindness = X é amável

[Kindness] has not been a leading concept in the normal School practices, neither in the classroom nor in external spaces. Neither has [respect] for the student regarding them as a Person of equal status as the adult, despite the current code of ethics that

[3] Definition of [amável] provided online by the Oxford Dictionaries.

regulates the set of rights and duties assigned to them while attending public school. This absence is one of the consequences of the centuries-old paternalistic ideology that views the child/young person not as a full citizen but as a project of a person whose construction will benefit from a robust disciplinarian attitude on the part of the adults responsible for their education.

This perspective also minimizes the importance of the child's mental states resulting from the impact of potential stress and/or hard circumstances, as if these did not have an equal or greater impact on a child than they do on an adult. It is expected, for example, that a child whose emotional world is destroyed by the separation of parents or is shaken by the painful permanent and irreversible absence of a loved one, or by the absence of a parental figure,[4] will regain their disposition to learn after a few psychological sessions. As if by being a child, the negative existential impacts, or the deep negative impacts caused by adults' life options would be more superficial and easily healable. After a brief period of tolerance regarding the eventually less successful performances by distressed children, the relentless standardized metrics will be applied, ignoring depressed emotional states or lack of motivation to learn.

Kindness and respect for the student's person are not incorporated as they should be in the institution's regular good practices, emerging occasionally thanks to individual goodwill, as in the case of an educational aide, responsible for managing a hallway with an average frequency of 6 classes of 20 students every 50 min. She made it a point to know not only the names of every student, but also to know at the beginning of the school year about their birthday so that on that day, she could wish them a happy birthday. This gentle natural kindness was reciprocated with the friendship and affection of all students, without exception. True kindness is made up of small gestures, spontaneous acts of pure generosity, which are not done to be politically correct or to get a better performance evaluation. These small great gestures make the School less gray, more humanized, and, therefore, closer to the heart of those attending it.

In the traditional school framework, a somehow neutral and formal pedagogical discourse is assumed as normal. But this neutral speech can quickly assume a reproachful tone or turn into an admonition if any of the conventional procedures is not observed or in the case of poor performance. Yet, the teacher's kindness towards their students is fundamental for successful learning and essential for a positive mind set. By fostering a learning environment where s/he is the solicitous element and facilitator of the magic of learning, the teacher illuminates possible tracks for research/study, encouraging students to seek answers/solutions to questions, without the fear of making mistakes, without fear of being penalized for poor performance, without the fear of making a poor, ridiculous image of themselves.

However, we have not yet reached that point and a long road must be traveled so that School can free itself from an ideology that associates learning, not with the wonder of discovering, but with an arduous, arid, and exhausting effort to incorporate

[4] I recall the distressful state of mind of a 12-year-old girl, living with two mothers and wanting to know who her biological father was. "Everybody has got a father. I just want to know who my biological father is, please."

often uninteresting subjects, useless information, an effort carried out under less favorable conditions and constant stress. This perception of arduousness, boredom and even pain kills in many students the enthusiasm for learning, the desire to open windows to unknown realities.

The perspective that learning is just an arduous joyless process, achieved only with personal sacrifice, is similar to the view that associates work with a burden that humans must carry on their back, a heavy toll, and not as a fantastic generative power that distinguishes human beings from all other species and whose driving force, throughout times, is responsible for human evolution and reality as we know it today. Yes, under hard conditions learning and/or working can be mentally and/or physically challenging but even so you can be passionate about it.

To conclude the importance of the concept of kindness in building interpersonal relationships, we would like to mention that in a social context, either physical or virtual, guided by the natural and widespread practice of kindness, there is very little or no place for the destructive phenomenon of bullying and other forms of violence. More than multiple formative actions aiming to eliminate these highly damaging social behaviors, we are convinced that the exercise of kindness in the school environment, its daily practice as a component of education to be incorporated by all, will substantially reduce the emergence of destructive behavioral patterns between peers or the violent acts that tv news report from schools all over the world.

Pope Francis, in Fratelli Tutti, recalled the importance of kindness in human relationships [19]:

> [….] an act of kindness, a concern not to offend by word or deed, a readiness to alleviate their burdens. It involves speaking words of comfort, strength, consolation and encouragement and not words that demean, sadden, anger or show scorn.

Acknowledging the role of true kindness as a harmonizer of human relationships, Francis points out: [ibidem].

> […] Often nowadays we find neither the time nor the energy to stop and be kind to others, to say "excuse me", "pardon me", "thank you". Yet every now and then, miraculously, a kind person appears and is willing to set everything else aside in order to show interest, to give the gift of a smile, to speak a word of encouragement, to listen amid general indifference. If we make a daily effort to do exactly this, we can create a healthy social atmosphere in which misunderstandings can be overcome and conflict forestalled. Kindness ought to be cultivated; it is no superficial bourgeois virtue. Precisely because it entails esteem and respect for others, once kindness becomes a culture within society it transforms lifestyles, relationships and the ways ideas are discussed and compared. Kindness facilitates the quest for consensus; it opens new paths where hostility and conflict would burn all bridges.

On the primordial role played by affection in social relations and particularly in the school environment, we recall the words of a student interview [16]:

> What I like most about school is being among colleagues whom I consider friends. Attending a school, in the company of people we like, allows us to face classes and tests more positively. By this, I mean that socializing and forming friendships is beneficial both for our self-esteem and for our education, as we do not feel we have to 'go through' this without support[5]

[5] English translation from the Portuguese original.

We will conclude the present chapter by addressing the third category proposed in Allardt's model.

2.2.5 [Being]: Shaping the Social Actor

In infancy and early childhood individual identity [Being] is defined by a set of features that consubstantiates kinship relationships established within the family: "child of X", "sibling of Y", "grandchild of Z".... However, as human beings progressively occupy a place in the larger community, assuming different roles, the semantic substance that defines this identity progressively incorporates the traits that define their different roles in society, translated into their concrete role/actions in each of these domains [10, 11].

The first social role in the larger community is that of student, incorporated upon starting school. [Being] will progress through a lattice of roles and social networks [10, 11]. The school is the place where [Being] is first functionally realized. The learner, therefore, in addition to occupying their place in the institution as a student, must have the opportunity to give their opinion on all matters that concern them individually or that concern the educational community to which they belong. The school must create the conditions for children, from an early age, to have the opportunity to be active members, expressing their opinions or making proposals that are **really** taken into consideration.

By giving voice to children, society is fostering their participation and development in society and by acknowledging their being, their status as full-fledged citizens, expanding the quality and robustness of democracy.

References

1. Convention on the Rights of the Child, United Nations (1989) https://www.ohchr.org/en/instruments-mechanisms/instruments/convention-rights-child
2. Cohen, Howard (1982). Equal Rights for Children. Law and Philosophy 1 (1):159–162.
3. Second Treatise On Government—John Locke Release Date: April 22, 2003 [eBook #7370] [Most recently updated: December 25, 2021]. Produced by: Dave Gowan and Chuck Greif https://www.gutenberg.org/files/7370/7370-h/7370-h.htm
4. Nishiyama, K. (2025) Children, Democracy, and Education A Deliberative Reconsideration. SUNY Press
5. Schapiro, Tamar (2003). Childhood and Personhood. Arizona Law Review 575 45:575–594.
6. Lyubomirsky, S., King, L., & Diener, E. (2005). The Benefits of Frequent Positive Affect: Does Happiness Lead to Success? *Psychological Bulletin, 131*(6), 803–855. https://doi.org/10.1037/0033-2909.131.6.803
7. Berscheid, E. (2003). The human's greatest strength: Other humans. In L. G. Aspinwall & U. M. Staudinger (Eds.), A psychology of human strengths: Fundamental questions and future directions for a positive psychology (pp. 37– 47). Washington, DC: American Psychological Association.
8. Allardt, Erik, (1993), Having, Loving, Being: An Alternative to the Swedish Model of Welfare Research', in Martha Nussbaum, and Amartya Sen (eds), *The Quality of Life* (Oxford, 1993;

online edn, Oxford Academic, 1 Nov. 2003), https://doi.org/10.1093/0198287976.003.0008, accessed 17 Nov. 2024.

9. Konu,A., Rimpelä, W., Well-being in schools: a conceptual model, *Health Promotion International*, Volume 17, Issue 1, March 2002, Pages 79–87, https://doi.org/10.1093/heapro/17.1.79
10. Ferreira, Maria Isabel Aldinhas (2007) On Meaning: Individuation and Identity: The Definition of a World View. (2007) PhD Thesis. Faculty of Letters. University of Lisbon. Lisbon Portugal
11. Ferreira, M.I.A., (2011) On Meaning: Individuation and Identity. Cambridge Publishers. London
12. Ferreira, M.I.A. (2011) Interactive Bodies: The Semiosis of Architectural Forms. Biosemiotics, Springer January 2011, volume 1, number 1. https://doi.org/10.1007/s12304-011-9126-0
13. https://www.publico.pt/2007/03/19/portugal/noticia/modernizacao-das-escolas-secundarias-ate-2015-vai-custar-mil-milhoes-de-euros-1288728
14. Munkegaard School Project Website https://arnejacobsen.com/works/munkegaard-school/
15. Azevedo, J. (2024) Modo de Produção da Exclusão Escolar. Olhar a escola a partir dos excluídos. Fundação Manuel Leão
16. Sol, https://sol.sapo.pt/2024/12/10/alunos-ha-muita-coisa-que-esta-mal-no-metodo-de-ensino-portugues/ (10 Dez 2024)
17. https://www.google.pt/search?q=the+role+of+kindness+at+school&sca_esv=41165fec71bc817d&sxsrf=ADLYWILPDuf6AtxbA2eMqkEkmIVbcgFz6A%3A1732709775655&source=hp&ei=jw1HZ6CfJaPW7M8P1
18. https://classful.com/16-reasons-why-kindness-is-important-at-school/
19. Encyclical Letter Fratelli Tutti of the Holy Father Francis on Fraternity and Social Friendship. https://www.vatican.va/content/francesco/en/encyclicals/documents/papa-francesco_20201003_enciclica-fratelli-tutti.html

Chapter 3
Working Paper #2 Immigration, Inclusion, and Linguistic and Cultural Diversity

1 Going Beyond the Concept of Tolerance

The concepts of equity and inclusion have become key topics of educational science discourse and policy priority across Europe [1: 17] to respond to the increasing cultural and linguistic diversity within society and consequently in the school landscape [2].

Working Paper #2 will briefly address the complexity posed by the linguistic and ethnic variety resulting from the migration waves targeting EU memberstates and that presently define rich but complex multicultural scenarios in the school learning environment, by taking a particular Portuguese case study. With this, we aim to contribute to "going beyond the limited perspective of tolerance' towards mutual knowledge and respect [3, p. 14] leading to what we believe will be a rich multicultural and multilinguistic social mosaic where even the tiniest communities will find their place to harmoniously evolve while contributing to the collective well-being of an economically stronger and more resourceful society.

2 Societies Are Evolving Social Organisms

Societies are living organisms that, like all living organisms, are in permanent configuration/reconfiguration. Throughout human history, population movements, motivated by environmental, social, economic, or political factors, have been a constant, determining new demographic configurations, defining and redefining the ethnic, linguistic, and cultural composition of communities, states, and regions, favoring the evolution and transformation of societies and even the formation and/or disappearance of civilizations.

According to [4], migratory populations can comprehend:

M. I. A. Ferreira, *Transitions of the Immutable: Reflections on Public Education in the 21st Century*, Studies in Systems, Decision and Control 625,
https://doi.org/10.1007/978-3-032-02809-9_3

– Asylum seekers: people who move across borders in search of protection […] who have applied for protection as a refugee and are awaiting the determination of their status.

– Refugees: any person who, due to 'a well-founded fear of being persecuted for reasons of race, religion, nationality, membership of a particular social group or political opinion, is outside the country of their nationality and is unable or, owing to such fear, is unwilling to avail themselves of the protection of that country.

– Migrants: individuals who have resided in a foreign country for more than one year irrespective of the causes, voluntary or involuntary, and the means, regular or irregular, used to migrate.

In this text, this distinction is not considered, with the three cases being simply addressed as migratory movements that give rise to temporary or permanent new settlements in a targeted region or country.

Europe, as we know it today, is the product of multiple inward and outward movements that displaced millions of people at different moments of our common historical narrative.[1]

War, religious and/or political persecutions, autocratic and repressive regimes, scarcity of natural or structural resources, climate crisis, or simply the search for new life opportunities have led and continue to lead millions to move, leaving their place of origin, their typical habitat, their modus vivendi, often risking their very existence, in search of a better life context for themselves and their families.

It is very educational to keep this fact in mind when hosting newcomers and remember all the times European populations were compelled to leave their homes, even their continent, e.g. the European diaspora to America, at different historical times, and follow difficult tracks towards a life that was envisaged as much more promising than the one lived so far.

In the case of the Portuguese diaspora, we can recall, for instance, the large and mostly illegal migratory movements, in the 60s of the twentieth century, towards France—a long way walking from Portugal to the Pyrenees, the hardship of crossing those mountains, running away from an impoverished country, from a dictatorship that had also enrolled in a colonial overseas youth devastating war. A long exhausting walk feeling hungry, tired and cold but always hoping to arrive at a better future. And not even the generally harsh work conditions finally found on arrival and the miserable life in the bidonville, swept away the expectation of a better future (Figs. 1 and 2).

Immigrants are frequently felt as "strange" by their hosting communities, the feeling of strangeness increasing when racial, cultural and linguistic features are strongly distinct. In 21 lessons for the 21st Century [6], Harari refers that it takes decades to achieve something like complete integration even if this is successful. On this topic, the Common Basic Principles (CBP) for Immigrant Integration Policy in the EU [7] highlight the complexity of the process:

> Integration is a dynamic, long-term, and continuous two-way process of mutual accommodation… It demands the participation not only of immigrants and their descendants but of every resident. (CBP. 1)

[1] Cf. on this topic: https://en.wikipedia.org/wiki/European_emigration.

Fig. 1 Carrying all the belongings in a card bag-aspects of the Portuguese migration to France in the sixties (twentieth century) [5]. *Credits* Gerard Bloncourt [2]

Fig. 2 Living in a bidonville (Paris suburbs)—Portuguese migrants in France 20th, in the 60s. *Credits* Gerard Bloncourt [2]

For this dynamics to be productive and successful two conditions must be fulfilled:

1. Immigrant populations are heartfelt willing to be a part of the host society by abiding to their laws and knowing and respecting the host's cultural values while contributing with their work to the individual and collective wealth and prosperity.
   1.2 The host country is a friendly but legally aware receptionist in that it develops equity policy frameworks capable of assisting and integrating newcomers but simultaneously legally controls and fights all forms of traffic of human beings that feed illegal immigration.
   1.2 The same legal frameworks regulate the work conditions and the daily life of native and immigrant workers.
   1.3 The host facilitates the integration by providing immigrants the same dignified working conditions as those available for native members.

According to [8] in the Common Basic Principles for Immigrant Integration Policy in the EU, we find the three cornerstones of multiculturalism: the right of all peoples to maintain their cultures, the right to participate fully in the life of the larger society, and the obligation for all groups (both dominant and non-dominant) to engage in a process of mutual change.

The process of integrating into a society is complex and somehow tricky. The European Union acknowledges this complexity that involves a process of mutual accommodation by the host and by the newcomers.

One of the difficulties in this process is the pre-assumptions held: on the one hand the host community rightly expects the newcomers to abide by their values and rules [5: CBP 2 and CBP 4] respecting and adapting to their new social environment. But hosts also expect newcomers to be less salient as possible in terms of cultural differences, speaking the host's language and exhibiting identical patterns of social behavior while integrating the community's workforce. This signifies complete acculturation and the vanishing of their original culture. Ignoring the potential contribute of migration to strengthen economy and social welfare, host populations frequently somehow feel they are being kind by letting these new elements integrate their social environment, by giving them the chance to share their social organization and national wealth and that this openness and "generosity" should be corresponded by a certain degree of gratitude, translated in the extreme respect towards the host's values and lifestyles.

Among the 11 principles of the Common Basic Principles for Immigrant Integration Policy in the EU, is the right to maintain one's original culture and religion (CBP 8):

> The practice of diverse cultures and religions is guaranteed under the Charter of Fundamental Rights and must be safeguarded, unless practices conflict with other inviolable European rights or with national law.

However, according to [8] there seems to be evidence that some host communities make a clear distinction between the right of immigrants to maintain their cultures in private (i.e., in their families and ethnic groups) but that does not include its expression or reflection in public domains, such as the educational or the work settings. Apart from being a cultural waste, not acknowledging their relevance and significance and not giving space to other cultures is one of the deep roots of the most violent ethnic conflicts taking place in some regions of Europe, involving the second and third generation of immigrants. Not having attained proper integration in the host society, not having found their role in the building of a common social edifice, not having their culture acknowledged they burst in acts of violence towards a society they don't recognize as theirs.

The fallacy of the secular character of the state and of its institutions (a republican claim) is responsible for ignoring that cultural and religious values have to be socially acknowledged and respected and that the role of the secular state is not to erase them, their symbolism, from school or other institutions, but to acknowledge they are present, embodied by those attending it, guaranteeing, this way, that all of them in their specificity and diversity will be equally respected. This means that people belonging to different cultures will be entitled to wear their typical garments, except for those that prevent a quick visual recognition of the user's face; that the cultural and religious calendar will be respected, at work and at school as the ones of the host community; that the school as it happens in airports will provide a dignified space—though not religiously symbolically marked-suitable for prayer; that the culturally determined dietary restrictions will be taken into consideration as any other.

As Berry [8] points out adaptation will be more successful in culturally diverse societies when individuals engage with both their heritage culture and the larger society.

Cultural diversity fosters economic, social and cultural wealth. When real integration is achieved and individuals have the chance to engage evenly both with their heritage culture and that of the host society, participating with their individual cultural richness in the daily life of the large community, this society becomes stronger and more prosperous.

Newcomers should be entitled to fully participate as culturally DISTINCT but NOT DIFFERENT people in the daily life of the society and its institutions, have their main celebrations acknowledged in the official holiday calendar and have their decorations illuminating the nights of the cities in a common shared celebration.

3 Immigration in Numbers

It is consensual that migratory movements are determined by two types of factors that, although distinct, complement each other: the so-called "push factors"—the negative factors in the country/region of origin that drive emigration, and the "pull factors", the attractive factors in the destination country/region.

According to the Eurostat, by the 1st of January 2023, 27.3 million citizens of non-member countries were residing in an EU Member State, representing 6.1% of the EU population [9].

Foreign-born population by country of birth, 1 January 2023

	Total		Born in another EU Member State		Born in a non-EU country	
	(thousand)	(% of the population)	(thousand)	(% of the population)	(thousand)	(% of the population)
Belgium	2,246.9	19.1	937.8	8.0	1,309.1	11.1
Bulgaria (¹)	168.6	2.6	57.7	0.9	110.9	1.7
Czechia	764.2	7.1	139.4	1.3	624.7	5.8
Denmark	804.1	13.6	263.5	4.4	540.6	9.1
Germany (³)	16,476.4	19.5	6,274.3	7.4	10,202.1	12.1
Estonia	234.7	17.2	26.6	1.9	208.1	15.2
Ireland	1,150.6	21.8	347.8	6.6	802.8	15.2
Greece	1,173.2	11.3	235.2	2.3	938.1	9.0
Spain	8,204.2	17.1	1,579.5	3.3	6,624.7	13.8
France (²)	8,942.1	13.1	1,988.9	2.9	6,953.2	10.2
Croatia	490.6	12.7	65.3	1.7	425.3	11.0
Italy	6,417.2	10.9	1,563.2	2.6	4,854.1	8.2
Cyprus	208.8	22.7	97.2	10.6	111.7	12.1
Latvia	241.1	12.8	22.1	1.2	219.0	11.6
Lithuania	231.8	8.1	18.6	0.7	213.2	7.5
Luxembourg	333.1	50.4	219.3	33.2	113.8	17.2
Hungary (¹)	644.5	6.7	342.2	3.6	302.3	3.1
Malta	153.4	28.3	40.3	7.4	113.1	20.9
Netherlands	2,777.0	15.6	747.9	4.2	2,029.1	11.4
Austria	1,963.3	21.6	863.0	9.5	1,100.3	12.1
Poland (¹)	933.1	2.5	231.5	0.6	701.6	1.9
Portugal	1,683.8	16.1	378.2	3.6	1,305.6	12.5
Romania (³)	531.0	2.8	202.0	1.1	329.1	1.7
Slovenia	309.3	14.6	62.3	2.9	247.0	11.7
Slovakia	213.2	3.9	156.3	2.9	56.9	1.0
Finland	461.2	8.3	131.0	2.4	330.2	5.9
Sweden	2,144.3	20.4	547.0	5.2	1,597.3	15.2
Iceland	82.2	21.2	53.0	13.7	29.3	7.5
Liechtenstein	13.6	34.4	7.2	18.0	6.5	16.3
Norway	965.5	17.6	368.4	6.7	597.1	10.9
Switzerland	2,662.0	30.2	1,458.5	16.5	1,203.5	13.7

Note: The individual values do not add up to the total due to rounding and the exclusion of the 'unknown' country of birth group from the table.
Bulgaria, Poland, Portugal, Slovakia, Finland, Sweden and Liechtenstein did not include
refugees from Ukraine who benefit from temporary protection in their population and migration statistics.
(¹) Break in series
(²) Provisional
(³) Estimate.
Source: Eurostat (online data code: migr_pop3ctb)

eurostat

Fig. 3 Distribution of the migrant population according to the country of origin, in 2023 [9]

Figure 3 shows the distribution of the migrant population according to the country of origin, in 2023 (source Eurostat).

In Portugal immigration has been rising steadily as Fig. 4 illustrates. Still according to the statistics in 2022, 118,000 immigrants entered Portugal, the highest number on record.

The five nationalities with the most immigrants residing in Portugal in 2020 were, in order, Brazil, the United Kingdom, Cape Verde, Romania, and Ukraine.

According to NUTS II,[2] [11] the total number of migrant children and young people enrolled in primary and secondary education in Portugal (mainland) has increased from 49,669 in the 2016/17 school year to 142.760 in 2023/2024.

As we can conclude by observing the data in Fig. 5, immigrant density is higher in the Lisbon region (47.098), North (30.009) and the Algarve region (12.319). This is easily explained as Lisbon Metropolitan Area, which includes the capital and the

[2] The NUTS nomenclature is a hierarchical system that divides the territory of the EU into different levels: NUTS I, NUTS II, and NUTS III, each corresponding to a specific geographical scale. NUTS II is used for the application of regional policies, corresponding to administrative regions or groupings of regions.

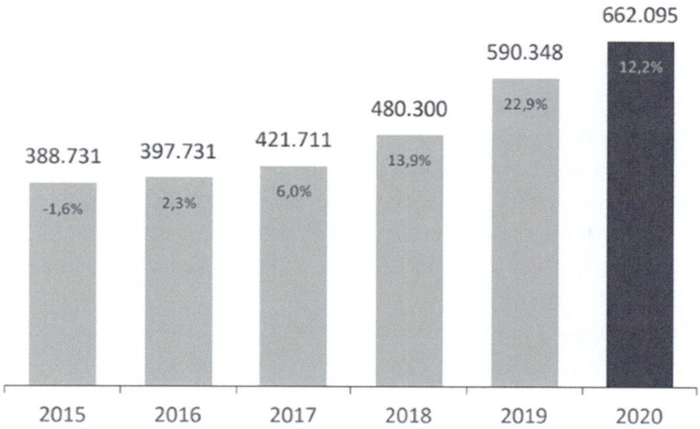

Fig. 4 Immigration to Portugal in the last decade [10]. *Source* Immigration, borders, and Asylum Report 2020, p. 20—Serviço de Estrangeiros e Fronteiras

suburban areas gravitating around it, is the biggest job provider, the same happening with the North region, starring Porto, the second biggest city and the Algarve the main region in terms of tourism. However, this density is not directly proportional in terms of the percentage of foreign students at school. The Algarve having the highest percentage of foreign students in public schools (19.5%), followed by the Lisbon Area (16.6%) and the Setubal Region (15%). This may lead us to conclude that in these three regions we have the presence of complete families and not just single individuals.

NUTS II	Estrangeiros	% do total de alunos por região
Norte	30 009	7,3%
Centro	19 142	10,4%
Oeste e Vale do Tejo	11 229	11,4%
Grande Lisboa	47 098	16,6%
Península de Setúbal	16 556	15,0%
Alentejo	3 990	7,5%
Algarve	12 319	19,5%
RAA	635	2,0%
RAM	1 782	5,9%
Total	**142 760**	**11,3%**

Nota: As fontes de dados sobre os alunos estrangeiros na RAA e na RAM são, respetivamente, a Secretaria Regional de Educação, Cultura e Desporto dos Açores (SRECD) e o Observatório de Educação da Região Autónoma da Madeira (OERAM).

Fonte: CNE, a partir de DGEEC, 2024

Fig. 5 Distribution of immigrants according to regions and corresponding immigrant children's school attendance percentage

4 The School—The Hub of Integration of Immigrant Populations

For immigrant children and adolescents, school is the most important arena to experience real social, linguistic and cultural participation in the host society. Participating in the school dynamics and thus learning together in the multicultural classes, where the natives interact with students coming from different parts of the globe is essential to foster inclusion and erase in the native population any eventual feeling of strangeness towards newcomers.

In fact, school plays a fundamental role in what concerns hosting and subsequent inclusion of migrant populations. This is referred in [3: 4-I].

> Schools have the potential to be real hubs of integration for children and their families. Increasing the participation of migrant children and children with a migrant background in ECEC while ensuring that such programs are equipped to serve culturally and linguistically diverse children can have a strong positive effect on their future educational attainment, including learning the host's language, and on the integration of their parents and families at large. A school environment that involves communities, support services and parents can be particularly beneficial for children with a migrant background. [....] Fighting segregation in schools and fostering interactions between migrant and native children, is also crucial to making our education system more inclusive. By teaching democracy, citizenship and critical thinking skills, schools play an important role in preventing young people from being attracted to violent extremist ideologies, organisations and movements.

The migratory flows that brought multicultural coexistence to Portuguese society, namely to the school's learning environment, required the adaptation of curricula to that new reality. Like the Portuguese children and teens, all immigrant students have the right and an obligation to receive educational training in the local school. Since 2006, the school curriculum contemplates the subject Portuguese as a Non-Native Language (PLNM).[3] This has been the response of the educational system, at the basic and secondary levels of schooling, for students whose mother language is not Portuguese and who show a linguistic proficiency of level A1, A2, or B1, according to the Common European Framework of Reference for Languages (CEFR) [12] or who have not had Portuguese as their language of schooling. In the graph in Fig. 6, we can observe the number of students that have attended PLNM classes in the school year 2022/23, according to their nationalities and school's level (Básico or Secundário).[4]

On the other hand, the Portuguese Language Welcome Courses (PLA),[5] created by Ordinance No. 183/2020 of August 5, following the previous Courses of Portuguese for Speakers of Other Languages (PFOL), are a response to the needs of adult migrants to meet what is legally defined in the immigration policy of EU member states:

---

[3] Português Língua Não Materna.

[4] Básico: from the 1st up to the 9th grade and Secundário from the 10th up to the 12th grade.

[5] Português Língua de Acolhimento.

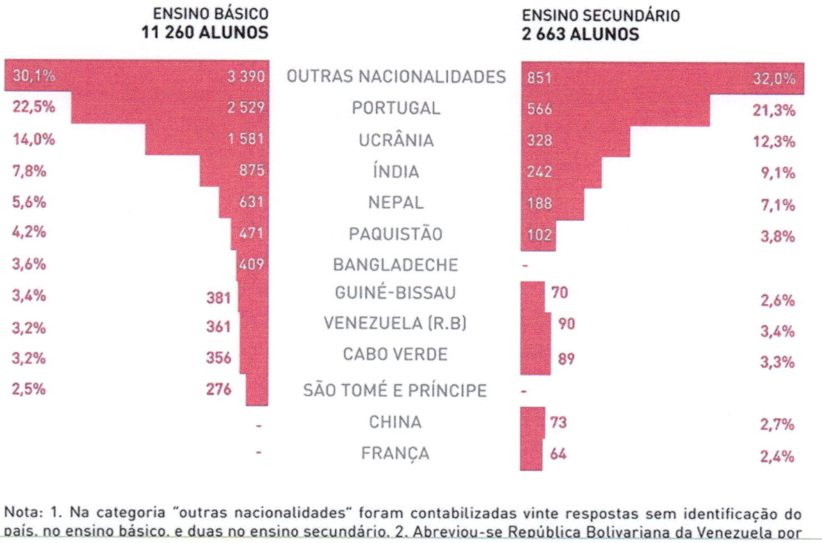

Fig. 6 Distribution of the students enrolled in PLNM in Public Schools in Basico and Secondary, according to the most representative nationalities (school year 2022/23—data from the mainland) [13]

> Basic knowledge of the host society's language, history, and institutions is indispensable to integration; enabling immigrants to acquire this basic knowledge is essential to sucessful integration. (CBP2)

This knowledge is not only a legal requirement, but it is above all a communicative necessity. Newcomers learn first the so-called basic "language for survival", progressively broadening their communicative competence to correctly function at work and comfortably interact in their daily routines. This can be exemplified by the efforts to learn Portuguese made by Uber drivers, coming from India and Pakistan or in the case of those working in the delivery services by Glovo or Uber Eats.

Typically, immigrant children learn the new language at school in the PLNM classes and also in the interaction with their peers, they usually learn it nearly effortlessly, because they are children.

The success of the inclusion process, translated in the bigger or smaller degree of immigrant students' participation in the school dynamics, depends on the conditions for participation set by the system.

Even if, as [8] points out, it might be a challenge for immigrant students to participate on an equal basis because of factors related to acculturative stress, such participation depends most of all on school systems' ability to meet immigrants students' needs in a way that allows them to experience inclusion.

5 The Limits of My Language Are the Limits of My World[6]

Languages embody distinct views of the world and unique ways of being [14, 15]. Since infancy, human beings conceptualize their being in a physical and social/cultural world first through the interactions within the family circle and subsequently interacting with other social peers, in a never ending process that progresses along the individual's lifetime, exponentially increasing their linguistic and communicative proficiency [14, 15].

According to Unesco [16] there are about 7000 languages spoken around the world today. But this linguistic diversity has been increasingly threatened as more and more languages disappear at an alarming rate. And when a language disappears, it takes with it an entire cultural and intellectual heritage.

Linguistic diversity should be treasured firstly, because human conceptualizing and symbolic capacity, that languages reflect, is—together with the generative capacity for work—an essential intrinsically anthropological asset and, secondly, because it is the key for humanity's shared knowledge.

Languages are in permanent evolution thanks to the human beings that use them. Changes in demography, namely migratory flows have always been one of the deep causes for that evolution.[7] As [17] refers Language is a function of the history of its speakers and the varied situations in which speakers may find themselves.

When human beings, especially children, depart from their original linguistic homeplace, the host country's language progressively takes the place of their mother tongue, just kept, at best, for interactions at home. To preserve an individual linguistic wealth that will ultimately also be a societal wealth, inclusion policies should not only foster the development of a communicative competence in the host country's language, but also nurture, in the case of children and young people, the development of the original linguistic competence in their mother tongue, and the associated historical/cultural knowledge, by integrating these subjects as options in the school curriculum.[8]

Amidst the challenges posed by multilingualism and the unprecedented explosion of cultures, in a world that is increasingly interdependent, the school acts not only as a fundamental facilitator of harmonious integration building bridges between the newcomers linguistic and cultural realities and those of the host society but also as a creator of opportunities for plurilingual education fostering the possibilities for participation in global conversations, work, and cultures.

The benefits of a plurilingual approach were already emphasized, in a different context, by the Common European Framework of Reference for Languages [12]:

> In an intercultural approach, it is a central objective of language education to promote the favourable development of the learner's whole personality and sense of identity in response to the enriching experience of otherness in language and culture (page:1) [....] The language

---

[6] Personal translation of Wittgenstein's quote: *Die Grenzen meiner Sprache bedeuten die Grenzen meiner Welt.*

[7] e.g., Romance and Germanic languages.

[8] Teachers on these subjects to be recruited through the local embassies of immigrant communities.

> learner becomes plurilingual and develops interculturality. The linguistic and cultural competences in respect of each language are modified by knowledge of the other and contribute to intercultural awareness, skills and know-how. They enable the individual to develop an enriched, more complex personality and an enhanced capacity for further language learning and greater openness to new cultural experiences. (page 43)

In a world that is increasingly interdependent, plurilingual education creates further opportunity for participation in global conversations, work, and cultures. In this context, there is an obvious value to develop a communicative competence on different languages.

6 Nurturing Linguistic and Cultural Diversity at School

6.1 A Case Study

The social and pedagogical challenges posed by a new demographic context are highlighted in Reimagining Our Future Together [18], which also identifies the pathway to follow to build a peaceful and harmonious society where everyone finds their place and role:

> The increasing mobility of human beings across the world, whether through choice or forced displacement, has created new pedagogical realities that bring the cultural and racial diversity of the world directly into classrooms and educational settings. Teachers are working in new environments with students who have diverse educational histories, languages and cultures. Pedagogies of respect, inclusion, belonging, peacebuilding, and conflict transformation go beyond merely acknowledging or tolerating difference. They must support students to sit side by side one another and work together. Education that allows young people to understand and link their pasts, presents and futures, analyze the inequalities that shape their experiences, stand up to exclusion and marginalization, is one of the best preparations for unknown futures. (page 53)

Within the Portuguese legal framework designed to address the present migratory phenomena and considering the limits of their autonomy, Portuguese public schools have been developing their own local initiatives to answer the eventual challenges posed by their specific regional demographic realities.

The case-study presented describes some of the initiatives taken in the course of the school years 2019/2020, 2022/2023 and 2023/2024 at a public school in the metropolitan area of Lisbon. In 2023/24 this school (5th–9th grades) had a universe of about 1000 students of which 300 were immigrants, distributed by the existing classes, according to their level of knowledge and age.

To address the multicultural diversity,[9] present at that school, three types of initiatives/educational actions could be identified in the time span 2019–2024:

[9] Besides the PLNM classes targeting the cohort.

(a) Multicultural manifestations—Random initiatives developed by teachers along the school year, involving immigrant students and their native peers, within the scope of the School's Yearly Plan of Activities—Plano Anual de Actividades.
(b) A global project—Acronym IDILICA (Integrating Linguistic and Cultural Diversity)—proposed by Profs Isabel Ferreira and Helena Pinto to the pedagogical council of the school and to the board of direction in the year 2023/2024.
(c) Multicultural Classroom Practice—developed at Class 8 of the 5th grade involving the Citizenship and the English classes, during the school year 2022/2023.

We will present them following this order.

6.2 Multicultural Manifestations in the School Yearly Plan of Activities

Schools generally have a yearly plan, comprehending optional educational activities to be realized during that time frame. These activities have the goal of developing student's competences in domains as literacies—e,g., reading and/or writing contests, dictation competitions, reading marathons; film sessions and film quizzes, literary contests; science and maths–quizzes and competitions; sports–basketball and football or handball tournaments; artistic performances; exhibitions….; school visits to places of interest. This type of activities frequently marks a particular day, e.g., Earth Day, Poetry Day…. Or marks a festive time—e.g., Christmas Time.

However, these yearly plans hardly ever contemplate significant dates other than those of the prevalent culture, despite the presence in the school setting of migrant populations belonging to other cultures.

An example is the case of the Orthodox Christmas celebrated on the 7th of January by Ucranian, Russian, Greek….Christians and largely ignored at the Portuguese public schools. Regrettably students from those regions see the local typical Christmas decorations being removed from the school setting, at the latest, on the 6th of January (traditionally assumed as the date ending the Christmas season), with no festive decoration being left to mark the next day and the date that is meaningful to them.

Though an overarching integrating strategy is completely absent in what concerns the newcomers's culture, attempts to give visibility, at school, to migrant students' cultures take place randomly.

This is the case, for instance, of the celebration of the New Lunar Year, in 2020, some weeks before the closure caused by the Covid pandemic. On this day, special sessions dedicated to the significance of the date in Chinese culture and its main cultural expressions (gastronomic including) were led by Chinese students and attended by several classes of the 5th and 6th grades (Fig. 7).

Fig. 7 Knowing about the traditions around the New Lunar Year Festival in China (session led by Chinese students). *Source* https://bibliotecasescolaresaen.blogspot.com/2020/01/ano-novo-chines-na-eb-d-antonio-da-costa.html

As it was already mentioned, the very unkind social behavior of ignoring other peoples' meaningful festivities is the result of the republican fallacy of the secular state, which advocates that governance should not favor any form of religious expression in public institutions, namely at school. By intentionally ignoring these cultural manifestations, the official context devalues the importance of the cultural narratives grounding them, limiting their expression by migrant students and consequently denying them equal status as peers.

With the aim of taking small steps towards a perspective of inclusiveness that, even at school, is not yet consensual, the plan of activities for the years 2022/2023 and 2023/2024 comprehended a couple of initiatives giving salience to the linguistic and culturally distinctive features present in that school context (Fig. 8).

According to the Yearly Activity Plan then defined, the first of these actions took place within the scope of the European Day of Languages—26th of September,[10] an initiative of the Council of Europe, that has been celebrating it since 2001. The Council assumes the rationale for this celebration in the website with the following statement:[11]

---

[10] https://edl.ecml.at/Home/tabid/1455/language/en-GB/Default.aspx.

[11] https://edl.ecml.at/Home/tabid/1455/language/en-GB/Default.aspx.

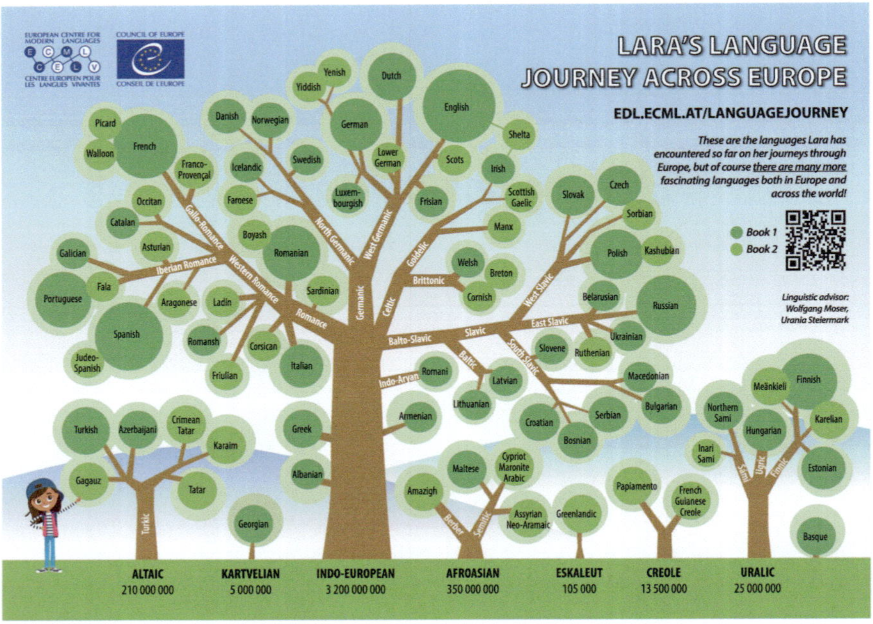

Fig. 8 Acknowledging European linguistic diversity—language tree poster. Credits to the Council of Europe

> Throughout Europe, 700 million Europeans are represented in the Council of Europe's 46 member states and all are encouraged to discover more languages at any age, as part of or alongside their studies. This stems from the Council of Europe's conviction that linguistic diversity is a tool for achieving greater intercultural understanding and a key element in the rich cultural heritage of our continent. The Council of Europe, therefore, promotes plurilingualism throughout Europe.

Therefore, at least once a year, European students are invited to acknowledge both the importance of speaking other languages than their mother tongue, the richness of European linguistic diversity and the kinship ties linking languages between themselves, as Figs. 9, 10 and 11 illustrate.

Apart from all the activities that the program for celebrating the European Day of Languages each year, always, suggests, a series of language workshops were organized in the school years 2022/23 and 2023/2024, aiming to make linguistic diversity more visible by bringing it closer to all the students. Each of these language workshops had a duration of 50 min and addressed an audience of about 20 students.

In each of the workshops, the hosts—one or two migrant students—presented their country of origin and main traditions/celebrations, using powerpoint, canvas or other digital templates (Fig. 12).

This brief introduction was done in English, in the case the hosts could not deliver it (yet) in Portuguese. Then an oral interplay between the migrant student leading the workshop and the students in the audience, followed. This interaction was the

Fig. 9 What languages can I speak?—a personal reflection

Fig. 10 What languages can I speak?—a personal and collective reflection

main substance of the individual workshops dedicated respectively to Ucranian, Moldavian, Russian, Hindi, Urdu and Mandarin (Fig. 13).

The student(s) hosting the session addressed the audience presenting and practicing in chorus and individually the basic structures and vocabulary associated to essential communicative functions[12] as

- Greeting
- Asking and giving information about oneself (Name and Age).

[12] The workshop structure and the way it should run was designed by Maria Isabel Aldinhas Ferreira and practiced with each of the migrant students, beforehand.

Fig. 11 A K-12 class with their teacher waiting for the language workshop to start

Fig. 12 The session's hosts getting ready to start

All the students involved, either those presenting the sessions or the classes participating in them, enjoyed being at the workshop, learning how to interact with others in a language they did not know.

Fig. 13 Andrei teaching the class how to write and pronounce some Russian words

6.3 The IDILICA Project

The IDILICA project (Integrar a Diversidade Linguística e Cultural) a proposal presented by Isabel Ferreira and Helena Pinto is an example of how difficult it is for a project to strive and succeed when the local institutional support fails. The project was presented to the Pedagogic Council and to the Direction Board of the school and approved in the year 2023/24. It aimed to contribute to the cultural and linguistic integration of the school's immigrant population through a series of coordinated actions involving different stakeholders: the authors of the project, the students, the immigrant students' parents or educators, the teachers and staff, the school's direction, the school's pedagogic council, representatives from the municipality, representatives from the embassies or consular organs of the original countries of the populations attending the school. As stated in the initial proposal, the main objective of the project was:

1. To contribute to the harmonic inclusion of migrant students in the school's context and dynamics by nurturing their cultural and linguistic heritage.

   This goal would be sustained by a series of coordinated actions, involving different stakeholders, namely:

   1.1. Identifying and quantifying the cultural diversity present at school.
   1.2. Identifying newcomers' expectations (students and parents) relatively to the new school environment and have their feedback, in regular meetings, (3 times a year) on how the adaptation is progressing and the main constraints found.
   1.3. Promoting opportunities for newcomers to participate in the school dynamics, namely through the election of national representatives—ambassadors.
   1.4. Creating opportunities for sharing the richness of linguistic and cultural diversity, fostering the interest and respect of national students relatively to the cultural reality of their new peers. An example of this were the language workshops led by immigrant students in years 2022/23 and 2023/2024.
   1.5. Scheduling the main celebrations of the cultural universes present at school in the school's celebrations calendar/timeline.
   1.6. Developing conditions that allow the young migrant populations to proceed with the learning of their mother tongue and historic and cultural narratives of their country of origin. This would be possible by integrating this topic, as optional, in the public school's curriculum. Qualified teachers would be recruited with the collaboration of the existing diplomatic venues.
   1.7. Creation of small clubs or, even better, departments according to the linguistic varieties present, e.g., Department of Slavic Languages, Department of Arabic Languages…. These small nuclei, created and organized by some of the school's language teachers in collaboration with the regional consular entities (that could, among other things, provide visual material and feed with updated cultural information) and parents would be run by

students, assisted by a teacher. These nuclei would also be the dynamo of social events to take place at culturally significant dates.

1.8. Supporting the integration of student's parents by providing the teaching of Portuguese to adult immigrants and the formal certification of their competences.

However, though having been approved, the project did not find the minimal conditions to be implemented except for the signaling of relevant dates and also the intercultural dynamics that took place in the classroom practice of a few classes. One of those is described next.

6.4 *Multicultural Classroom Practice in Class 5º8*

In the school year 2022/2023, class 8 of the 5th grade was a heterogeneous class, of 20 students, constituted by a Portuguese majority, plus two Ukrainian, two Bengali, two Nepali, one Brazilian, an English born Pakistani and a Russian.

Given this linguistic variety and the fact that most of the migrant students were just starting to learn Portuguese, the schooling language used in classes was English, as at least its basic formulations could be understood by all.

Two of the works developed by these students in the English and in the Citizenship classes were posted in the school's website with the following introduction:[13]

> At a time when School increasingly defines itself as a mosaic of cultures, it is essential for everyone to become aware of the importance of the linguistic and cultural legacy they carry, in the definition of their own identity and in the building of a common reality.

Class 5º8 is a small instance of this growing multiculturalism. Throughout the 1st semester, in Citizenship and English classes, and through various activities, students talked about their origins and culture and became aware of the richness that this variety brings to the class. "Citizens of the World" marks this diversity that converges into a common reality that "Let's Celebrate"[14] reflects".

"Citizens of the World", Fig. 14, highlights the fact that coming from different parts of the planet a small number converged at a particular time and space and became peers. The small pink ballons indicated the geographical location and also the number of students having that origin, as the balloon on Portugal, Fig. 15, indicates by showing the number 14.

The texts and illustrations were produced and chosen by the individual students. Surfacing each of the locations opens to the individual personal profiles. Proper name, age and some favourite activities or other preferences were the type of phrases chosen to complete the small fact file (Figs. 16, 17 and 18).

[13] /https://www.aeen.pt/2023/03/07/citizens-of-the-world-lets-celebrate.

[14] https://padlet.com/isabelaldinhas/let-s-celebrate-2pc92cm479jpcy7m.

Fig. 14 Citizens of the world

Fig. 15 Short personal profiles 1

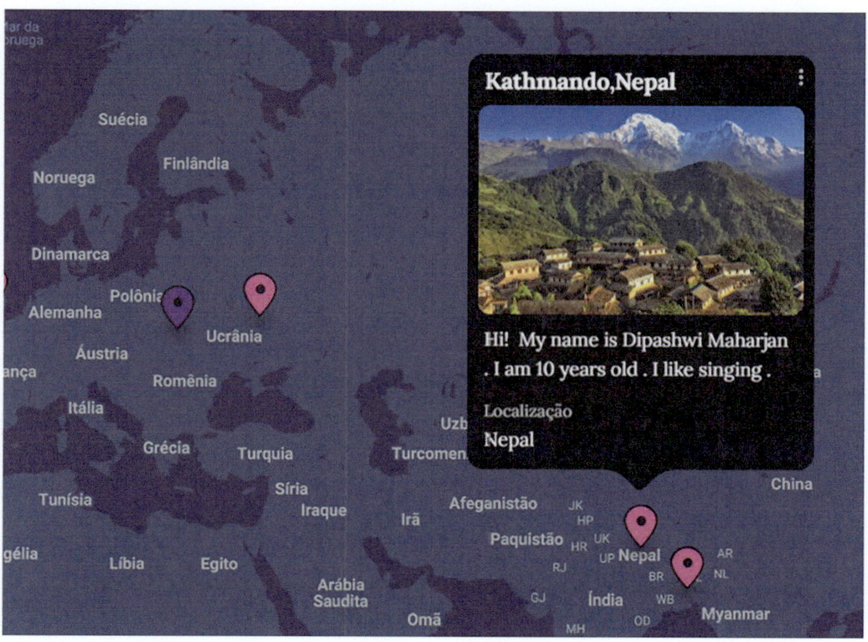

Fig. 16 Short personal profiles 2

Fig. 17 Short personal profiles 3

Knowing about people's special dates and eventually partying at those times is a good way to build cohesion within a group and foster friendship ties. The Celebrations' Calendar, in the form of a timeline that's starts in September and finishes in June highlights significant individual dates, such as one's birthday and cultural relevant times (Fig. 19).

Here again the texts and illustrations were produced by the students and are part of a process of knowing each other and enjoying learning and being together.

For the sake of space the long timeline is reproduced in several fragments (Figs. 19, 20, 21, 22, 23 and 24).

The terms community, diversity and inclusion animate official educational policy discourse and are present not only in the education policy-statements of governing states and their educational institutions, but also in the discourse of organisations as the OECD, UNESCO, UNICEF or World Bank. And, yet, these concepts, these

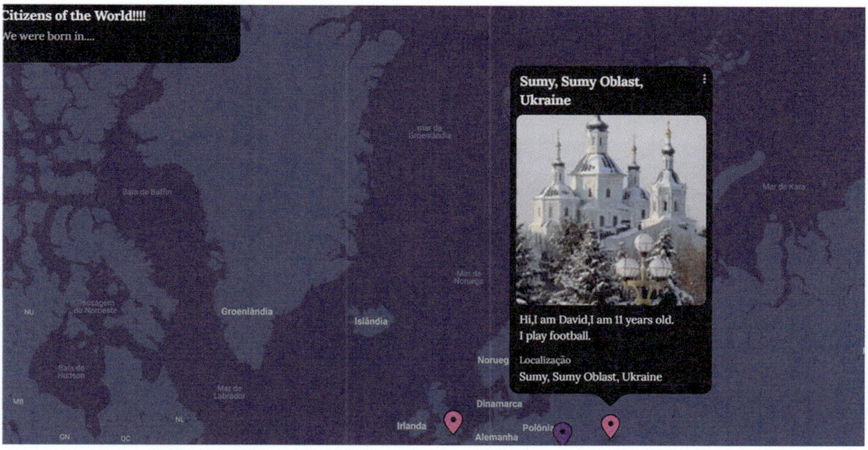

Fig. 18 Short personal profiles 4

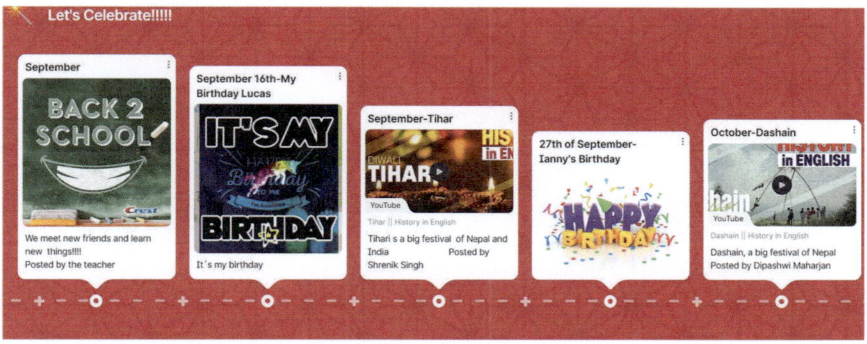

Fig. 19 Let's celebrate—5º8 calendar of festive events 1

Fig. 20 Let's celebrate—5º8 calendar of festive events 2

Fig. 21 Let's celebrate—5º8 calendar of festive events 3

Fig. 22 Let's celebrate—5º8 calendar of festive events 4

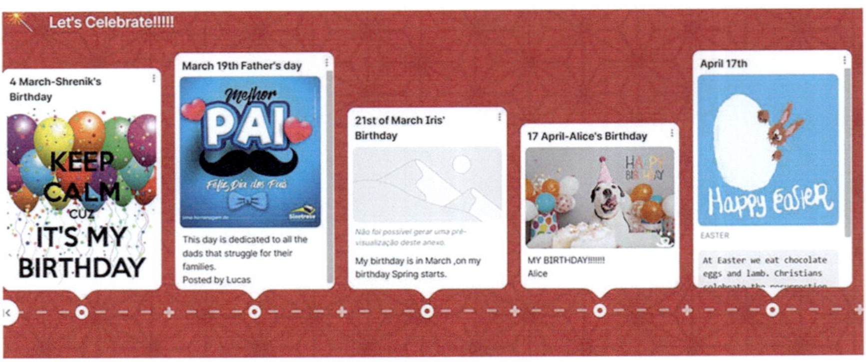

Fig. 23 Let's celebrate—5º8 calendar of festive events 5

Fig. 24 Let's celebrate—5º8 calendar of festive events 6

terms, only make sense if/when translated in real practice, by coherent strategies or at least by such small acts as these described above.

The activities here described were small steps that don't leave a trail in the essential immutable character of the institution but that may leave a seed of transformation in the heart of each student.

References

1. UNESCO 2017—Annual Report. https://unesdoc.unesco.org/ark:/48223/pf0000261971
2. "Reimagining Our Futures Together: A New Social Contract for Education", UNESCO 2021. ISBN 978-92-3-100478-0. https://doi.org/10.54675/ASRB4722
3. Communication from the Commission to the European Parliament, The Council, The European Economic and Social Committee and the Committee of the Regions Action plan on Integration and Inclusion 2021–2027 COM/2020/758 final. https://eur-lex.europa.eu/legal-content/EN/TXT/?uri=CELEX%3A52020DC0758
4. https://home-affairs.ec.europa.eu/policies/migration-and-asylum/legal-migration-resettlement-and-integration/integration/action-plan-integration-and-inclusion_en
5. Photos by Gerard Boncourt. https://www.youtube.com/watch?v=v9xQ-9oVscY
6. Harari, Y.N. (2019) 21 Lessons for the 21st Century. Vintage Publishing
7. https://migrant-integration.ec.europa.eu/library-document/common-basic-principles-immigrant-integration-policy-eu_en
8. Berry, J.W. Comparative analysis of Canadian multiculturalism policy and the multiculturalism policies of other countries. https://doi.org/10.11621/pir.2016.0101. https://psychologyinrussia.com/volumes/?article=4621
9. https://ec.europa.eu/eurostat/statisticsexplained/index.php?title=EU_population_diversity_by_citizenship_and_country_of_birth
10. Immigration, Borders, and Asylum Report 2020, p. 20—Serviço de Estrangeiros e Fronteiras. https://www.sef.pt/pt/Documents/RIFA2022%20vF2a.pdf
11. https://migrant-integration.ec.europa.eu/news/portugal-oecd-report-highlights-increasing-migrant-student-numbers_en
12. Common European Framework of Reference for Languages (CEFR) https://rm.coe.int/1680459f97

13. Conselho Nacional da Educação. Relatório "Estado da Educação em 2023". https://www.cnedu.pt/content/EE2023/Versao_Integral/EE2023.pdf. 1.ª Edição Dezembro de 2024 ISSN 1647-8541 ISSN Digital 2976-0267
14. Ferreira, Maria Isabel Aldinhas (2007) On Meaning: Individuation and Identity: The Definition of a World View. (2007) PhD Thesis. Faculty of Letters. University of Lisbon. Lisbon Portugal
15. Ferreira, M.I.A. (2011) On Meaning: Individuation and Identity. Cambridge Publishers. London
16. UNESCO 2022 https://www.unesco.org/en/articles/why-mother-language-based-education-essential
17. Wilbert Heeringa, Charlotte Gooskens, Vincent J. van Heuven, Comparing Germanic, Romance and Slavic: Relationships among linguistic distances, Lingua, Volume 287, 2023, 103512, ISSN 0024-3841, https://doi.org/10.1016/j.lingua.2023.103512. https://www.sciencedirect.com/science/article/pii/S0024384123000360
18. Reimagining our Future Together—A New Social Contract for Education https://unesdoc.unesco.org/ark:/48223/pf0000379707.locale=en

Chapter 4
Working Paper #3 Technological Revolution, Pedagogical Innovation, and Ethics

How we learn must be determined by why and what we learn
Reimagining Our Futures Together—A new social contract for education Chapter 3 I Pedagogies of cooperation and solidarity 50.

1 The New Technological Artifacts and Their Relationship with the User

The revolution, introduced by Information and Communication Technologies, the Internet, IoT, and the latest developments in robotics and Artificial Intelligence, is one of the dynamos of transformation in modern society, determining different modes of production and distribution; altering the nature of work and the way we relate to its processes; influencing how we interact, thus shaping behaviors, routines, and lifestyles, favoring the creation of a hybrid existential reality, where the physical and the virtual interpenetrate, overlap, and sometimes confuse each other.

The 1980s and 1990s were fundamental milestones in the ongoing technological revolution, enabling the exponential growth of computer developments recorded up to that point. This present technological revolution has been significantly contributed to by (i) the miniaturization of electronic components and the increase in processing capacity that led to the popularization of personal computers and the emergence of the first mobile phones; (ii) the advent of the World Wide Web (WWW), developed by Tim Berners-Lee, making the internet accessible to the public and the creation of browsers like Netscape Navigator and later Internet Explorer, which allowed the expansion of web use; (iii) and currently, the development of intelligent applications—AIs (Artificial Intelligent Systems)—in multiple domains.

At the beginning of the twenty-first century, broadband internet became widely available, allowing the multiplication of streaming services, online games, and other activities that required greater bandwidth. But it was the emergence of smartphones like the iPhone in 2007 that caused a massive transformation by allowing the average

M. I. A. Ferreira, *Transitions of the Immutable: Reflections on Public Education in the 21st Century*, Studies in Systems, Decision and Control 625,
https://doi.org/10.1007/978-3-032-02809-9_4

user immediate access to the internet, anywhere and at any time, thus access to a multitude of digital platforms. Consumers enthusiastically embraced the communicative and expressive potential offered by these new tools, introducing them uncritically into their lives, often in an excessive and alienating manner, even recording moments of their intimacy and personal life on the available virtual platforms, completely ignoring that, by doing so, they were entirely giving up their privacy and being unaware of how their personal data, preferences, and virtual journey fueled the monumental data storage business.

Smartphones, these fantastic little objects, computationally more complex and powerful than the computers that supported humans in their first trip to the Moon, have become inseparable companions of their users, almost an extension of their own bodies and minds, allowing permanent contact with others, research, and collection of any type of information, and have become guides and navigation aids in space and time. But the exponential increase in their capabilities and the range of their applications has turned what was meant to be occasional presence and interaction into a permanent and often dominant presence, regardless of the context of the lived experience. This demanding and almost cognizant presence progressively invaded the public and private spheres, becoming highly intrusive and destabilizing the desirable relational, communicative, and even affective environment inherent in human interaction in presence and socialization in a physical context. At a time when their presence is questioned in various environments, particularly in School, we must ask ourselves if prohibiting their use in certain contexts is really a good solution. Smart technologies have arrived, and it is up to humans to learn how to relate to them, using them to their benefit while preserving what is best in them—their humanity. It is at School that this learning should take place; it is up to School to educate.

2 Information and Communication Technologies in Public Schools in Portugal

In Portugal, the introduction of Information and Communication Technologies (ICT) in non-university public education began in the 1980s and has progressed to the present.

The MINERVA Project (Meios Informáticos no Ensino: Racionalização, Valorização, Actualização), [1] (1985–1994) was the first major national initiative, financed by the Ministry of Education, whose primary objectives were the training of teachers in primary and secondary education with a view to introducing new technologies in the teaching/learning process and in student assessment, as well as the inclusion of ICT teaching in curricula.

During the 1990s, schools continued to incorporate computers and the first technological tools in teaching, albeit in a limited way. The projects then underway were mainly aimed at equipping computer labs and providing basic training to teachers of various subjects on the use of information and computing technologies.

The Technological Education Plan (PTE)—approved in 2007 [2] is the largest national technological modernization program in Portuguese public schools and the one that had the greatest impact both in the short and long term. This plan allowed schools to be equipped with high-speed internet, as well as ensuring that all classrooms in the 2nd and 3rd cycles and secondary education were equipped with a computer and a projector. This fact considerably contributed to introducing the progressive diversification of strategies to be developed in the classes of various subjects and the use of digital resources, thus improving the dynamics of the teaching/learning process. At this stage, the ICT class became an integral part of the curriculum. The PTE also included the e-School program [3] which aimed to ensure universal access for primary school students to computing resources, guaranteeing their access to personal computers with educational content and generalizing the use of computers and the internet in early learning. To this end, a small laptop computer was developed by Intel, designed for children aged six to eleven years and equipped with educational software appropriate to the age and level of schooling of its young users. This computer was named 'Magalhães,' Fig. 1, as a metaphor for navigating the "seas" of the Web.

The mass distribution of these computers in 2009 (more than 500,000 according to the 2019 Education in Numbers report) [4], and the educational program implemented in the 1st cycle under the PTE, allowed students to enter the 2nd cycle already possessing basic skills in using this equipment, such as being able to conduct simple research, write a short text in Word, or make a basic presentation in PowerPoint. This

Fig. 1 Magalhães: navigating the WWW and developing skills at Primary School. Credits to: https://pplware.sapo.pt/informacao/magalhaes-o-pc-de-portugal-alem-fronteiras/

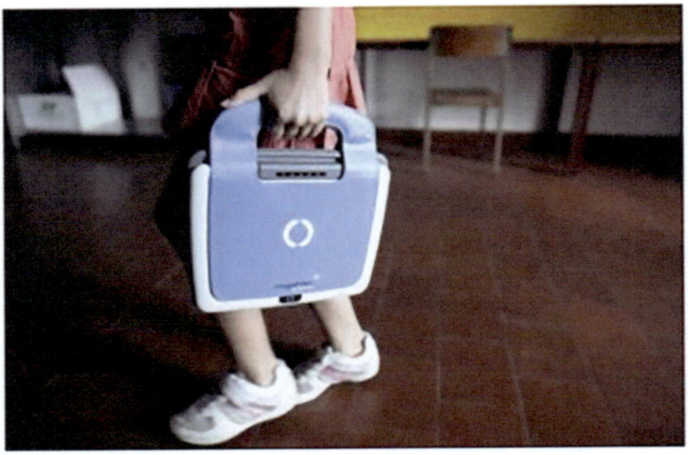

Fig. 2 The children's laptop to be used at school and at home

facilitated the application of these digital skills in the work developed in various 2nd cycle subjects. Students continued to demonstrate these skills during the period the e-School program was in effect, but ceased to do so after the program ended (Fig. 2).

In the narrative defined by this process of incorporating technology into the teaching–learning process within the public school system, we can identify two fundamental objectives: (i) ensuring universal access to the digital reality by providing the necessary means for such (computers with high-speed internet access), and (ii) promoting the literacy necessary for productive usage. Absent from this initial programmatic line was the concern with developing a critical mindset in users regarding this reality, an education in ethical terms, and the development of awareness that could be called "healthy" usage of digital platforms, particularly social networks.

The fact that educational and computing technologies showed a reasonable level of incorporation into teaching practices in 2020, both in terms of their application in student performance evaluation and classroom dynamics, was a decisive factor in how the public school system in Portugal was able to face the enormous challenges posed by the Covid pandemic, despite all difficulties, to both teachers and students and their families, during the months when in-person teaching transitioned to remote, and physical interaction gave way to digital interaction.

Moreover, the COVID-19 pandemic created the need for online interaction (synchronous classes) or offline (asynchronous classes), greatly accelerating the development and use of a plethora of digital resources and interaction platforms, e.g., Google Classroom, Google Meets, Teams, or Zoom, requiring all users, teachers, and students, to develop digital literacy capable of supporting cooperative work and efficient interaction in a virtual environment. The pandemic also exposed the existing inequalities among students, both in terms of possession of computer equipment and connectivity conditions: Students without a computer at home, or with only one computer for multiple users (parents working online, or siblings, in the case of families with several children), students with low-resolution connectivity, or without connectivity to any internet server. In these situations, it was often the smartphone, which practically all students already possessed or could access, that allowed the remote communication between students and the School.

With the aim of leveling the inequalities detected during the confinement period, giving all students identical conditions so they could recover learning that had been hindered and in order to "leave no one behind," the Digital School program was launched in November 2020 [5]. According to this program, all students should receive a kit consisting of a laptop, headphones, and a router with internet access paid for by the Ministry of Education, Fig. 3. This kit would remain in the possession of and be used by the students throughout their schooling, with the guardian required to return it at the end of the schooling period. An identical kit was distributed to all teachers who requested it.

According to empirical evidence in teaching practice, these computers, perhaps because it was not very practical to take them to school every day, were mainly used at home, being present in the classroom mainly during training moments, in the context of ICT classes, or in preparation for national digital exams, such as assessment tests in the 1st and 2nd cycles. Perhaps because their distribution did not produce the expected impact in terms of developing digital literacy, or due to the financial burden of their permanent connectivity, these devices were collected by educational services at the end of 2024, with their distribution expected to continue only for students with noted economic needs.

Fig. 3 Digital kit: so that no one was left behind

When discussing the digital/learning relationship, it is imperative to mention the presence of the smartphone, a device whose ownership and use have become widespread among practically all students from the age of 10 over the last decade and, as with adults, accompanies them throughout the day, consequently also being present at school and in the classroom. Regarding the prevalence of the smartphone and the preference for its use compared to the traditional laptop, Oluwafemi J. Sunday et al. [6] state:

> The global surge in use has led to an ever-increasing dependency on smartphones because of their embedded functionalities and portability factor. Some people, especially the younger generation, prefer a smartphone over a laptop and personal computers because of the various functionalities embedded in smartphones, such as entertainment, watching sports, online games, online banking, sending emails, chatting, agenda tracking, online shopping, social media, and the ability to complete homework. (2021:12)

3 Intelligent Digital Tools in School

3.1 The Case of the Smartphone[1]

According to a study conducted by Hubside.Store in 2018 [6] and referenced by SapoTek [7], 95% of Portuguese children aged 10 or older owned their own smartphone in 2021. The study was based on a sample collected through an online survey of 500 parents with children aged between 10 and 17 years. According to the evidence collected, 50% of children receive a smartphone at the age of 10, coinciding with their entry into the second cycle.

This fact is easily understandable, given that entering the 2nd cycle brings with it a series of transformations—(i) attending a new school, generally larger and with many more students, (ii) transitioning from a mono-teaching system—where basically a single teacher is responsible for the class, to a set of teachers (approximately eight) and to a diverse range of subjects with often distinct dynamics. It is, therefore, natural in this context that the mobile phone is perceived, both by guardians and children, as something that soothes the anxiety caused by the unknown and change, by allowing immediate contact in situations where this anxiety escalates, which can be simply caused by the child not being able to easily navigate that large space, locating the various points where activities should take place.

Thus, as has happened in other contexts, the smartphone has assumed a significant presence in the school environment. It can be said that probably the number of smartphones in a given school today roughly equals the number of young people and adults that populate it.

The Student Statute and School Ethics [8], which establishes the rights and duties of primary and secondary school students in public schools, and the commitment

[1] By the time this book is published, Portugal has banned the use of smartphones by students, from year 1 up to year 9.

of parents or guardians and other members of the educational community in their education and training from the 1st to the 12th year of compulsory schooling, sought in its articles to respond to the challenges posed by the presence of new technological tools and their various applications in an educational environment, defining the regulatory norms for their use. Thus, according to the legislation [9], Article 10, of the Duties of Students (subsections q, r, s, t), the student **must not**:

> q) transport any materials, technological equipment, instruments, or devices that could objectively disrupt the normal functioning of teaching activities or cause physical or psychological harm to students or any other member of the educational community;
>
> r) use any technological equipment, namely, mobile phones, equipment, software, or applications, in places where classes or other training activities or meetings of school bodies or structures are taking place, except when the use of any of the aforementioned means is directly related to the activities being developed and is expressly authorized by the teacher or the person responsible for directing or supervising the ongoing work or activities;
>
> s) capture sounds or images, namely, of teaching and non-teaching activities, without the prior authorization of the teachers, those responsible for directing the school, or supervising the ongoing work or activities, as well as, where applicable, any member of the school or educational community whose image may be recorded, even inadvertently;
>
> t) disseminate, at school or outside it, namely, via the Internet or other means of communication, sounds or images captured during teaching and non-teaching moments, without the authorization of the school director;

As we can see, the legislation seeks to prevent individual technological equipment from disrupting the school environment, whether in the classroom or outside of it, while simultaneously safeguarding situations that put at risk the right to data privacy and image protection for all members of the school community. However, the legislation leaves open the possibility that, under the guidance and supervision of the teacher, smartphones can be used in school, particularly in the classroom context, for the development of learning-related activities.

The internal regulations that schools, within their autonomy, drafted mostly followed what is legislated in the Student Statute and School Ethics until the end of the 2023/24 school year. According to a sample collected from different parts of the country (N = 10), the internal regulations of seven school clusters do not allow smartphones in the classroom; they must remain stored in backpacks, without sound or turned off. Their use, however, is permitted, even in the classroom, whenever requested and supervised by the teacher to perform certain tasks or activities. According to these seven regulations, outside the classroom, the student is free to use them, safeguarding the aspects referred to in subsections s) and t) of the Student Statute. In three of the school clusters in the collected sample, the presence of smartphones is prohibited in all school spaces and under all circumstances.

We are currently witnessing a growing questioning of the negative effects on the development and personality of children and young people due to excessive and prolonged exposure to electronic screens and interaction with devices such as smartphones. According to the 2019 World Health Organization report [10], it is recommended that children under one year of age not be exposed to screens and that children under five years old have a maximum of one hour of interaction with them, with parents always accompanying the viewing of content that is appropriate for the age level and always of high quality. In the context of the current growing global discussion, cf. [11, 12], prompted by scientific studies that have been developed [6], pointing to some of the harms of using smartphones by children: (i) vision problems in young children, disruption of their ability to concentrate, and even the danger of depersonalization; (ii) possible alienating effect and hindering concentration in class, causing sleep disturbances, anxiety and addiction, inhibiting in-person socialization in older children (2nd and 3rd cycles) and young people (secondary school and university students), it is natural that the Ministry of Education and Science in September 2024, issued a recommendation for schools to ban the use of mobile phones by students in any school space in the 1st, 2nd, and 3rd cycles, prohibiting their intrusive and potentially disruptive presence in the school [13].

Believing that the smartphone should not be handled by young children, given the recognized harms in terms of vision and mental structuring and that its presence in school brings no benefit until the end of the 1st cycle, a stage where literacy should indeed be developed using physical support—books—and where writing, calligraphy should be emphasized for its importance in neuronal development and fine motor skills, we advocate that from this stage onwards, it is up to the School [14] as the training entity par excellence, to make young users aware of the enormous potential in terms of learning of that small portable object, which students possess and use simply as a communication channel and socialization platform. In this way, developing the ability to learn how to learn and autonomy, it is also up to the School to raise awareness of the highly negative factors that improper or excessive use can have as a consequence.

As Growing up Digital, Australia (ibid.) states: "[...] schools require children to use digital devices, they also have a responsibility to educate the children about appropriate use of the devices and also to educate them in ways to disconnect from devices [...]". Therefore, it is the School's duty to prepare young citizens for the correct, healthy, and ethical use of new technologies, particularly Artificial Intelligence, so that their use can translate into well-being and individual and collective development.

The following chapter aims to show how some teaching/learning strategies using smartphones can be particularly motivating, even for less interested students, fostering more autonomous learning and a collaborative spirit, as well as facilitating the development of true digital literacy, where know-how about the functional use of equipment is accompanied by the development of a critical mindset regarding the nature of the content viewed and the conditions of its use. These types of strategies were developed in foreign language (English) classes, with a duration of 50 min, in five classes of about 20 students, K-12, over three school years, consistently showing very positive results in overall learning terms.

3.2 The Smartphone in the Classroom: A New Way to Learn How to Learn—The Experience of the Foreign Language Class

Sunday et al. [6] state that despite the problems related to excessive exposure of children and young people to mobile phone screens and extensive interaction in this digital environment, some research has been pointing to positive aspects of using these devices, not only in mobility or distance learning situations [15] but also in the classroom context, particularly in self-regulated learning [16], improving students' active participation in class dynamics, developing collaborative spirit and peer interaction [17] Wang, Shen, Novak, Pan (2009) and language learning [18].

Learning a foreign language occurs through the development of communicative competence in the target language, from a very early stage—Pre-A1 to proficiency in C1 and C2, as stipulated in the Common European Framework for Language Learning, Teaching, and Assessment (CEF) of 2001, updated in September 2023 [19]. The CEF functions as the European standard for defining, regulating, validating, and certifying all foreign language teaching/learning processes, and the relevance of its parameters has influenced foreign language didactics worldwide. The CEF aims for this learning, thus standardized, to be holistic, i.e., to develop equally in both reception and production modalities. As the CEF states (ibid.):

> The language learner/user's communicative language competence is activated in the performance of the various language activities, involving reception, production, interaction or mediation.

The foreign language classroom is a very dynamic communication space, where interaction with the teacher and peers plays a fundamental role. This communication generally takes place using the foreign language itself, depending on the complexity of the statements at the learners' level.

Currently, the demographics present in schools (both public and private) reflect large global migratory movements, being composed of a heterogeneous population concerning their geographical origin, and mainly their linguistic and cultural traditions.

In this multilingual learning context, where there is often no common language that enables communication between teacher/student or between peers, as in the case, for example, of Chinese students who only communicate in Mandarin or Cantonese, the smartphone is a fundamental tool to expediently overcome this situation without affecting the overall classroom dynamics. I also recall in this context the case of a refugee student who, not speaking English or Portuguese, approached the teacher during a class, showing her smartphone screen. It read: "Please, can I go out, I have got a huge headache."

Regarding general classroom activities, the smartphone can also be a tool that enhances student autonomy by allowing them to know the meaning of a new lexical unit, verify pronunciation, or search for the best grammatical form for a given statement, making the information consultation usually present in the traditional language dictionary more immediate, complete, and user-friendly.

Another domain where the smartphone has proven its potential as a learning tool is in digital gamification, allowing its introduction into the classroom dynamics in a quick, efficient, and highly motivating way.

Gamification aims to enhance learning by associating it with (i) a strongly interactive and playful component and (ii) the challenge. By requiring the attention and engagement of all students, it results in greater active participation, thus significantly contributing to better academic performance. As [20] states, technology-assisted gamification activities increase the learning curve.

The usage described below was always determined by one of the following objectives:

1. To consolidate a topic (vocabulary or structural-grammatical) presented during the class.
2. To check knowledge regarding previously worked topics.
3. To review topics where students were still experiencing difficulties.

The materials used were:

1. The router assigned to the teacher, to which all students connected to access the internet, thus ensuring a network with equal speed for everyone.
2. One smartphone per pair of students.
3. The projector and classroom screen.

Methodology

For this activity, the class organized itself into teams (pairs of students) to foster a collaborative spirit. Each pair used a single mobile phone to access the interactive game platform. The platform used was always Kahoot [21]. The pair decided which student would give the team its name. Figure 4 illustrates Kahoot in team mode, as projected on the screen and also on the individual mobile phones.

In the illustrative case, the vocabulary topic was "Daily Routine" and the structural/grammatical content was the Simple Present of action verbs.

In Fig. 5, we can see the board projected on the class screen and also on the teams' mobile phones during the game. The teams will have to choose the correct option within seconds, according to the given sentence:

Figure 6 shows the names of the three teams that respectively finished in 3rd, 2nd, and 1st place after around 20 situations and a healthy competition between teams.

Finally, we would like to mention what is, in our opinion, the most comprehensive activity from the perspective of language learning, involving the use of the smartphone and some of the applications available on it.

Project Works

These projects always involve a planning phase before execution and essentially encompass skills in the production of oral statements: speaking production and/or speaking interaction. They are the chosen activity to generally conclude teaching unit sequences, where one or more morphosyntactic structures and specific lexical domains must be used for oral communication purposes. The script that guides the

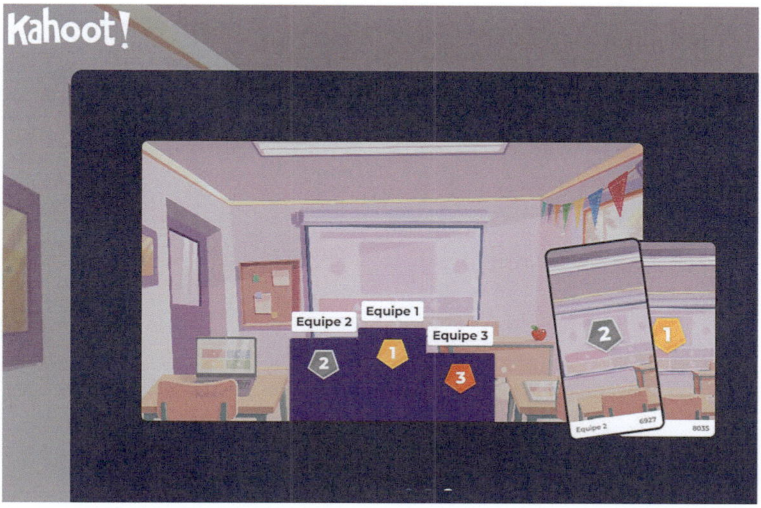

Fig. 4 Kahoot in team mode

Fig. 5 Choosing the correct option

execution and implementation of these projects is collectively constructed in class by the students under the teacher's supervision. The illustrative Figs. 7, 8, 9 and 10. are snapshots extracted from these short videos.

Project A: Interviewing Schoolmates

Video in which the student interviews a classmate of their choice (in English), addressing questions such as (i) school schedule, (ii) favorite subjects and why, (iii) things they do not like about school.

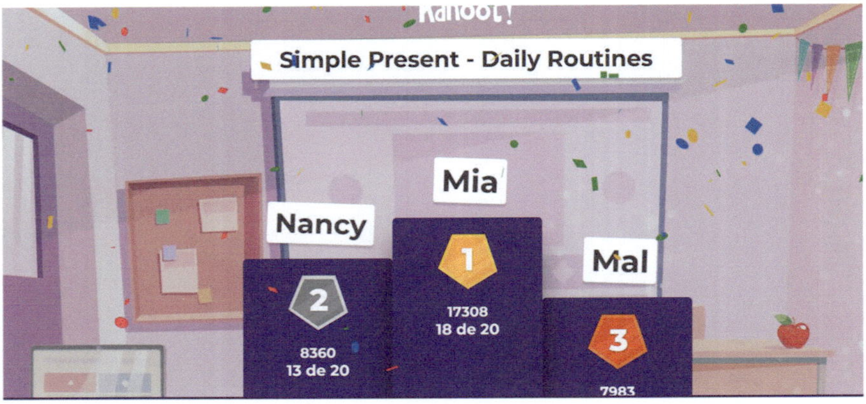

Fig. 6 Final podium

Fig. 7 Enzo interviewing Afonso in the school's library

The students choose the location, at school or outside of it, where they filmed this interview with a smartphone (often asking another classmate to assist with the filming).

Sometimes the students added written comments while editing the film that run simultaneously with the oral interview, as illustrated in Fig. 8, where the students joke about using a juice bottle as a microphone.

Frequently, the students chose an opening for their film as shown in Fig. 9 and accompanied it with a "technical sheet". It can be observed that the students used some Tik Tok applications, for example, to edit their video and select templates for the opening and technical sheet. In some cases, they even added a "behind the scenes" with comic mishaps and minor errors that occurred during filming, as shown in Fig. 9.

Fig. 8 Sofia Interviewing Helena

Fig. 9 Looking like a real film- abertura da entrevista de Catarina a Daniel

Project B: My Daily Routine on School Days

Individual video about their daily routine made by the student with their smartphone, with the level of detail, from waking up to going to sleep, left to the discretion of the students (Fig. 11).

Project C: Our School

Video in which the students, organized into small groups of 4 members, present their school, the classroom, and the various spaces, identifying the activities that usually take place there (Fig. 12).

Projects A, B, and C are simply examples of a student-centered linguistic practice focused on the production of oral statements and interaction with peers using

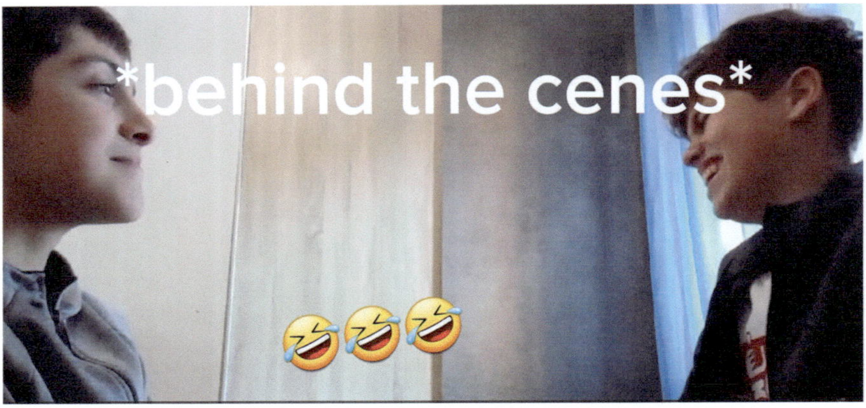

Fig. 10 Behind the scenes—Tiago's interview with Daniel shows how the students also had fun while doing the project (the misspelling was not intentionally corrected)

Fig. 11 Francisca's daily routine on school days

the English language (speaking production and speaking interaction), in which the smartphone played an essential role in filming and later in the editing that the students did autonomously. All the films were evaluated by the teacher according to predefined criteria, the discussion of which is beyond the scope of this article. The produced films were, with the authors' permission, viewed in English class by the respective classes. The four best videos in each category, elected by the participating classes and the involved teachers, were nominated for an Oscar in the categories of "Interview," "My Daily Routine," and "Our School," with the best nominee in each category receiving an Oscar at the end of the year in a Hollywood-style Oscar ceremony (Fig. 13).

Fig. 12 School facilities—the canteen (In Our School by Daniella, Benedita, Keylla and Marta)

Fig. 13 Slide from the ceremony presenting excerpts of the nominated films and awarding the Oscar in the relevant categories

4 Conclusion: Interactive Platforms, AI and Now What?

The WEF [22] emphasizes that the demands of this transforming society, characterized by what is conventionally referred to as the 4th Industrial Revolution, go far beyond typical literacies and curricular content, encompassing a set of skills that must necessarily be developed in and by the School: "Society requires learners to have a range of holistic skills to thrive in the modern world. These include creative, technology, innovation and interpersonal skills. And today, these skills and knowledge need to be acquired in a more accessible, personalized and active way than ever before." (2020: 8).

The School is not only the institution that can provide education on how to ethically handle digital/intelligent tools but simultaneously teach how to avoid the physical, mental and social risks that emerge from some of the forms of interaction that these tools make available.

The process of ideation and iteration that all innovation entails requires active, engaged, and participatory learning from the learner. To promote this type of learning, the traditional paradigm of a "process-based approach to learning" must be shifted to a "problem-based approach to learning," where students, though guided by the teacher, are agents of their own learning, defining through trial and error the path to solving problems as would happen in the real world, outside of school. This type of agency and autonomy is already a practice in informal learning so frequently initiated by young people on virtual platforms like Duolingo (for language learning) or in small demonstration videos made by YouTubers. As UNESCO [23] states: "Today's children are growing up in a world of abundant choice and personalized experiences enabled by technology. They have on-demand access to the videos and shows they'd like to watch, friends are one message away, and they can customize their mobile devices with the applications they find most useful, listed in the order that is most efficient for them to access." (2017: 5).

The Future Report commissioned by Google and delivered by Livity [24] interviewed 7000 students, from different countries, to find out their experience as users of digital tools. The report has revealed that in 2025 students are not passive users but are using intelligent tools to learn and to create.

According to this report, four out of ten students (40%) use AI tools nearly daily; 81% believe that AI enhances creativity and 65% think it helps generating new ideas. However, 28% refer that many schools don't allow the use of these tools.

By altering the type of intervention of the teacher, the classroom dynamics are substantially changed, centering it on the learner, making the learning process truly active and significantly improving its outcomes. As [25] points out (Figs. 14 and 15).

By giving the student control over their learning process, they become an active agent rather than simply a receptacle for content. This active role is highly appreciated by learners, who may even kindly refuse the teacher's help to experiment and test on their own. In this new environment, the teacher facilitates learning by opening "windows," pointing to possible resources, and alerting to potential problems.

This is simultaneously a hybrid learning process where experiences in real spaces and in the virtual/digital world converge, allowing for a multiplicity of contexts and forms of interaction that are recognized as highly motivating and thus conducive to successful learning. The consolidation of these learnings in project work can take the form of presentations in various digital applications, e.g., Canvas or Genially, video production, songs, exhibitions of various kinds, and community interventions.

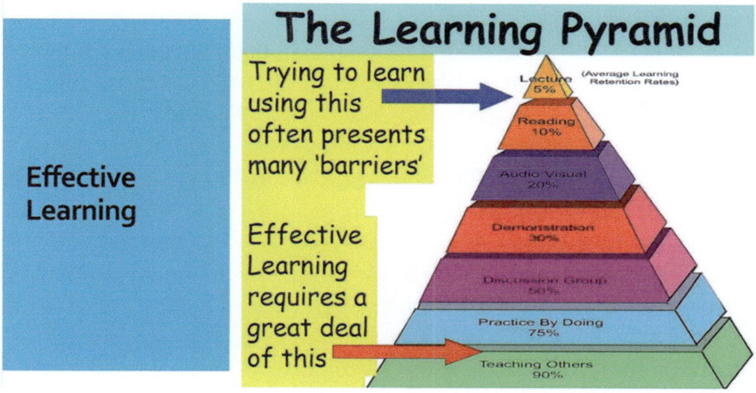

Fig. 14 Steps to effective learning. Credits to [25]

Fig. 15 The future report—How young people are using digital and intelligent tools. Credits to Google

The WEF [26] refers to the role played by these different technological applications: "Technologies designed to fully embrace opportunities for agency, guidance and creation while allowing playful interaction are some of the most powerful tools we have to support high-quality learning today."

The launch of ChatGPT in November 2022 [27], greatly expanded the capabilities of Artificial Intelligence and brought it closer to the common user. With its ability to respond to "prompts" by producing texts, videos, drawings, music, writing code… smart tools have significant implications at all levels of formal learning. In June 2023, UNESCO warned that Generative AI should be used with great caution in education, given the lack of regulation regarding its use in schools and the negative impacts already identified, such as data privacy issues, bias, manipulated and false news, among many others. According to UNESCO 2023, schools are not yet prepared either pedagogically or ethically for the regular use of Generative Artificial Intelligence.

It also suggests that it should not be used by students under the age of 13. The Recommendation also underscores the importance of the school's role in raising awareness and training students for the ethical and correct use of smart technologies.

Regarding the rapid proliferation of Generative AI applications, [28] states (2023:7): "The rush to embed AI in schools is driven less by explicit educational purposes and more by the visions and financial interests of the AI industry. AI technologies are extremely expensive to run, but AI in education is reckoned to be highly profitable."

On the other hand, reflecting on the potentially positive role of AI anticipating the student's eventual learning difficulties from the very start, so that the learning process can adapt to the student's characteristics, [29] points out that although further exploration is needed, the predictors enabled by this study could be used to improve personalization in edtech software, as well as to help teachers make better decisions about how they use classroom resources. The goal is to help teachers intervene at just the right time to encourage and celebrate progress.

In conclusion, in what concerns education, as in all the other domains, it is essential to understand that the ongoing technological revolution can bring huge benefits, once the potentially negative impacts are anticipated, so that they can be eliminated. In the school context, smart applications can be a working tool that enhances the cognitive abilities of its users and gives them greater autonomy in terms of learning; however, they will never replace the role of interaction with the teacher and peers, which, complemented by a set of smart tools, will constitute the most favorable context for learning. It will be up to the teacher, the ultimate trainer, to be able to guide their students in this process of learning to learn, supported by various forms of literacy and guided by a critical vision and ethical principles.

References

1. https://pt.slideshare.net/slideshow/relatrio-do-projecto-minerva/39813
2. https://diariodarepublica.pt/dr/detalhe/resolucao-conselho-ministros/137-2007-642198
3. Pereira, S., Pereira, L. (2011) Políticas tecnológicas educativas em Portugal: do Projecto Minerva à Iniciativa e-Escolinha. Congresso Nacional "Literacia, Media e Cidadania" 25–26 Março 2011, Braga, Universidade do Minho. Centro de Estudos de Comunicação e Sociedade ISBN 978-989-97244-1-9
4. A Educação em Números 2019 https://www.portugal.gov.pt/pt/gc22/comunicacao/comunicado?i=educacao-em-numeros-20192020
5. https://portugaldigital.gov.pt/formar-pessoas-para-o-digital/conhecer-a-escola-digital/
6. Oluwafemi J. Sunday, Olusola O. Adesope, Patricia L. Maarhuis, The effects of smartphone addiction on learning: A meta-analysis, in Computers in Human Behavior Reports. Volume 4, August–December 2021, 100114
7. https://www.sciencedirect.com/science/article/pii/S2451958821000622. Available online 24 June 2021

8. https://tek.sapo.pt/noticias/negocios/artigos/95-das-criancas-portuguesas-com-10-anos-ja-tem-um-smartphone,
9. Diário da República n.º 172/2012, Série I de 2012–09–05, páginas 5103–5119
10. World Health Organization (2019) https://www.who.int/news/item/24-04-2019-to-grow-up-healthy-children-need-to-sit-less-and-play-more
11. Graham, A, Sahlberg, P. (2021) Growing up digital Australia: phase 2 technical report https://apo.org.au/node/311990
12. https://eprints.lse.ac.uk/125554/1/Smartphone_policies_in_schools_Rahali_et_al_2024_002_pdf
13. https://www.portugal.gov.pt/download.ficheiros/ficheiro.aspx?v=%3D%3DBQAAAB%2BLCAAAAAAABAAzNDE2MgIA5VegJgUAAAA%3D
14. https://www.learning.com/blog/role-of-schools-online-safety/
15. C. Steel (2012) Fitting learning into life: Language students' perspectives on benefits of using mobile apps Ascilite (2012), pp. 875–880
16. L. Sha, C.K. Looi, W. Chen, B.H. Zhang(2012) Understanding mobile learning from the perspective of self-regulated learning. Journal of Computer Assisted Learning, 28 (4) (2012), pp. 366–378, https://doi.org/10.1111/j.1365
17. Wang M., Shen,R., Novak,D., Pan X., (2009) The impact of mobile learning on students' learning behaviours and performance: Report from a large blended classroom. British Journal of Educational Technology, 40 (4) (2009), pp. 673–695, https://doi.org/10.1111/j.1467-8535.2008.00846.x
18. Thornton P., Houser,C., (2005) Using mobile phones in English education in Japan Journal of Computer Assisted Learning, 21 (3) (2005), pp. 217–228, https://doi.org/10.1111/j.1365-2729.2005.00129.x
19. The Common European Framework for Language Learning, Teaching and Assessment', (CEF) de 2001, actualizado em Setembro de 2023. https://rm.coe.int/1680459f97
20. Kayımbaşıoğlu, D., Oktekinb B., Hacı, H., (2016) Integration of gamification technology in education. 12th International Conference on Application of Fuzzy Systems and Soft Computing, ICAFS 2016, 29–30 August 2016, Vienna, Austria https://pdf.sciencedirectassets.com/280203/1-s2.0-S1877050916X00282/1-s2.0-S1877050916326400/main.pdf?X-Amz-Security-Token=IQoJb3JpZ2luX2VjECIaCXVzLWVhc3QtMSJHMEUCIQDXWK3J2VEG9KWqX39Z9oJJTJ3rHEp
21. https://kahoot.com/
22. Schools of the Future Defining New Models of Education for the Fourth Industrial Revolution-World Economic Forum 2020. https://www.weforum.org/publications/schools-of-the-future-defining-new-models-of-education-for-the-fourth-industrial-revolution/
23. Artificial Intelligence in Education- Unesco. https://www.unesco.org/en/digital-education/artificial-intelligence?hub=32618. https://www.unicef.org/media/48581/file/SOWC_2017_ENG.pdf
24. The Future Report https://futurereport.eu/
25. https://pt.slideshare.net/slideshow/education-40-what-india-should-do/240607782
26. Learning through play: how schools can educate students through technology- World Economic Forum Annual Meeting—Jan 24, 2020. https://www.weforum.org/stories/2020/01/technology-education-edtech-play-learning/
27. https://openai.com/index/chatgpt/
28. Education in the age of artificial intelligence The UNESCO Courier October-December 2023 https://unesdoc.unesco.org/ark:/48223/pf0000387029_eng
29. Ge Gao, Amelia Leon, Andrea Jetten, Jasmine Turner, Husni Almoubayyed, Stephen Fancsali, Emma Brunskill (2025) Predicting Long-Term Student Outcomes from Short-Term EdTech Log Data. LAK '25: Proceedings of the 15th International Learning Analytics and Knowledge Conference 2025, https://arxiv.org/pdf/2412.15473

Chapter 5
Takeaways

Public schooling and education emerge as the pathway to preserve the fundamentals of humanness, hopefully leading to an harmonious individual and collective flourishing.

We remember, on this purpose, some fundamental insights of Reimagining Futures Our Futures Together: A New Social Contract for Education [2].

At this time, we would like to recall three fundamental concepts of Deleuze and Guattari's [1] perspectives on education-those of [Rizhome], [Assemblage] and [Becoming] that substantiate most of the problems we tried to address in the working papers above. The Deleuzian framework advocates for a student-centered experiment-driven approach to learning, one that highlights the dynamics of life in all its domains and fosters a process of transformation rather than static knowledge acquisition. By assuming the learning process as rhizomatic, i.e., as a decentralized horizontal network with multiple entries and exit points, those authors propose a model that might be more capable of reflecting the complex and interconnected way human beings presently learn. On the other hand, the concept of [Assemblage] highlights the varied ontological nature of the entities that compose the grid on which the very process of learning depends and how these entities connect and interact forming temporary functioning knowledge and communication lattices. The concept of [Becoming] highlights the dynamics of the teaching/learning process, one of constant transformation and change for both learners and teachers in their encounters with the unknown.

We would like to finish these reflections on public education in the 21st century, by recalling fundamental insights of Reimagining Futures Our Futures Together: A New Social Contract For Education [2].

> Education is the foundation for the renewal and transformation of our societies. […] a pillar of the 2030 Framework for Sustainable Development – an inclusive vision for humanity to advance well-being, justice, and peace for all, as well as sustainable relationships with the environment (Introduction-page 10)

M. I. A. Ferreira, *Transitions of the Immutable: Reflections on Public Education in the 21st Century*, Studies in Systems, Decision and Control 625,
https://doi.org/10.1007/978-3-032-02809-9_5

During the twentieth century, public education was essentially aimed at supporting national citizenship and development efforts. It primarily took the form of compulsory schooling for children and youth [...] emphasized values of individual success, national competition and economic development, to the detriment of solidarity, understanding our interdependencies, and caring for each other and the planet. [...] A new social contract for education is needed anchored in two foundational principles:

(1) the right to education

(2) a commitment to education as a public societal endeavour and a common good. (Pages 10–11)

References

1. For People-yet-to-Come: Deleuze, Guattari Politics and Education, Mathew Carlin and Jason Wall eds. (2014) Bloomsbury.
2. Reimagining Our Future Together—A New Social Contract for Education https://unesdoc.unesco.org/ark:/48223/pf0000379707.locale=en

MIX
Papier aus verantwortungsvollen Quellen
Paper from responsible sources
FSC® C105338

If you have any concerns about our products,
you can contact us on
ProductSafety@springernature.com

In case Publisher is established outside the EU,
the EU authorized representative is:
Springer Nature Customer Service Center GmbH
Europaplatz 3, 69115 Heidelberg, Germany

Printed by Libri Plureos GmbH
in Hamburg, Germany